Life

on a

Young

Planet

Life
on a
Young
Planet

THE FIRST

THREE BILLION

YEARS OF

EVOLUTION

ON EARTH

Andrew H. Knoll

PRINCETON UNIVERSITY PRESS

PRINCETON AND OXFORD

Fourth printing, and first paperback printing, for the Princeton Science
Library, 2005
Paperback ISBN 0-691-12029-3

The Library of Congress has cataloged the cloth edition of this book as follows
Knoll, Andrew H.
Life on a young planet : the first three billion years of evolution on earth /
Andrew H. Knoll.
p.   cm.
Includes bibliographical references (p.).
ISBN 0-691-00978-3 (alk. paper)
1. Life—Origin.   I. Title.

QH325 .K54 2003
576.8'3—dc21   2002035484

British Library Cataloging-in-Publication Data is available

This book has been composed in Palatino

Printed on acid-free paper. ∞

pup.princeton.edu

Printed in the United States of America

10 9 8 7 6 5 4

For my parents.

In nature and in nurture,

I was lucky.

Contents

Acknowledgments

THIS VOLUME distills the thoughts of a quarter century spent trying to understand the early history of life. I first entertained the idea of writing a book more than a decade ago, but fortunately got sidetracked. Children, research, and university responsibilities kept me from reconsidering such a plunge until the fall of 1998, when the alignment of growing kids, the end of my term as department chair, and a sabbatical leave convinced me that the time was right to attempt what for me was a new style of scholarship. Of course, my children weren't the only ones who had matured in the interim, and so whatever critical fate awaits this volume, I can honestly state that it is far better than it would have been had I completed it at first consideration.

For all the caricatures of scientists as creative loners, science is a richly social endeavor. Our worldviews evolve by reading the works of those who went before, by teaching and learning, through collaboration, conversation, and argument. The ideas and experiences related in the following pages owe much to others, many of them mentioned in one chapter or another. Elso Barghoorn guided my thesis research, providing me with opportunities, support, and a collegiality whose obvious asymmetry he never stressed. My graduate education was shaped as well by Ray Siever, Dick Holland, Steve Golubic, and the late Steve Gould and Bernie Kummel, all of whom seemed to spot in me a potential I would never have seen for myself.

The students and postdoctoral fellows in my lab have been a continuing source of joy and intellectual sustenance, and I thank them all. I have also benefited enormously from wonderful colleagues. My friends in biology and the Earth sciences at Harvard keep me ever on my toes.

In the world beyond Harvard Yard, I am particularly grateful for the friendship and intellectual stimulation of John Grotzinger, Sam Bowring, John Hayes, Malcolm Walter, Roger Summons, Keene Swett,

Yin Leiming, Misha Semikhatov, Misha Fedonkin, Volodya Sergeev, Gerard Germs, Stefan Bengtson, Simon Conway Morris, Brian Harland, Don Canfield, Ariel Anbar, Dave Des Marais, Ken Nealson, Sean Carroll, and my departed friends Zhang Yun, Gonzalo Vidal, and Preston Cloud.

Of course, books are not written in the office or in the field. They get finished in the upstairs study, on weeknights after the homework is done. Book writing, therefore, is very much a family affair. In a profession that commonly rewards obsession, my children have given me the gift of balance in life. And even by confessing that I lack words to articulate my gratitude, I risk trivializing the importance of my wife Marsha.

Much of my research over the years has been funded by the National Science Foundation and NASA, including the NASA Astrobiology Institute. I am grateful for their support. I also thank Dick Bambach, Susannah Porter, Don Canfield, Sean Carroll, Jack Repcheck, Kristen Gager, Lawrence Krauss, and Marsha Knoll for reading my draft manuscript and making many suggestions for improvement. John Bauld, Roger Buick, Stefan Bengtson, Martin Brasier, Birger Rasmussen, Shuhai Xiao, Richard Jenkins, Leonid Popov, Dave Bottjer, Steve Dornbos, Greg Wray, Andreas Teske, Susannah Porter, Bruce Lieberman, and Nick Butterfield provided some of the illustrations that leaven my text. Lastly, I thank Sam Elworthy and Princeton University Press for their unstinting support and confidence. Sam has been my Maxwell Perkins, improving every page of what follows.

Prologue

In his brief poem "When I heard the learn'd astronomer," Walt Whitman recounts an evening spent at a scientific lecture. Proofs and figures fill the hall, oppressively weighting the air,

> Til rising and gliding out I wander'd off by myself,
> In the mystical moist night-air, and from time to time,
> Look'd up in perfect silence at the stars.

Although written more than a century ago, Whitman's poem resonates with a surprisingly large contemporary audience. Earlier attempts to understand the universe and our place in it distilled nature's mystery into powerful narrative. Science, Whitman implies, replaces awe with statistics.

But does ignorance really outstrip understanding as the preferred route to wonder? As a paleontologist, I don't think so. To me, the scientific account of life's long history abounds in both narrative verve and mystery. Lucy's skull and diminutive bones, carefully displayed in a museum drawer, transport me to the warm African savanna where humanity took shape 3 million years ago. Dinosaurs take me back twenty to seventy times further, to Mesozoic forests patrolled by astonishing beasts—if I can't share the awe that *Tyrannosaurus* inspires in my son, it is, quite simply, a failing of maturity. Older yet are the trilobites, those joint-legged monarchs of the Cambrian seas, skittering around a tropical reef some 500 million years ago.

The fossils of animals, claimed by popular culture as much as by science, provide a biological chronicle of remarkable proportions. And yet, they record only the most recent chapters in Earth's immense evolutionary epic. The complete history of life ranges over four *billion* years, through alien worlds of sulfurous oceans beneath asphyxiating air, past iron-breathing bacteria and microscopic chimeras, to arrive at last at our

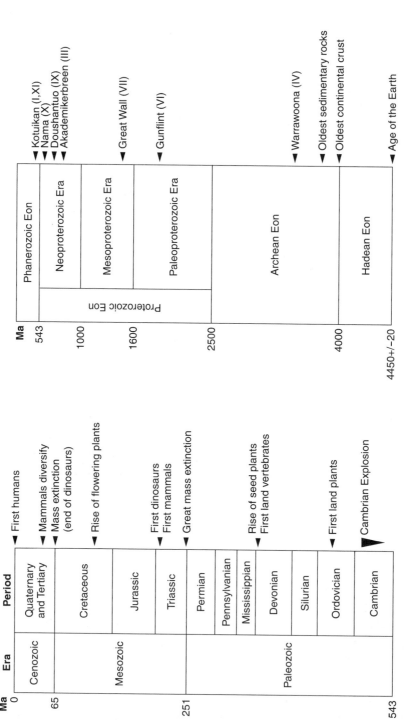

Figure P1. The geologic timescale, showing the time relationships of major events in Phanerozoic evolution and the Precambrian rock units discussed in this book. (Ma = million years before present)

familiar world of oxygen and ozone, forested valleys, and animals that swim, walk, and fly. Scheherazade could hardly have invented a more engaging tale.

Nor is the story complete in its current telling. It can't be, because each hard-won fact raises a new question. John Archibald Wheeler, one of the twentieth century's preeminent physicists, once remarked that we live on an island in a sea of ignorance. This metaphor comes with an insightful corollary: as the island grows, built piece after piece by the accumulation of knowledge, its shoreline—the interface between knowledge and uncertainty—expands proportionately. There is much we do not understand about the history of life, and the same will be true of our grandchildren. But, then, if we knew all there was to know, scientific interest would cease. Textbooks may portray science as a codification of facts, but it is really a disciplined way of asking about the unknown.

This, then, is a book about history—the history of life before the dinosaurs, before the trilobites, before animals of any kind. My story begins with the initial diversification of animals in Cambrian seas. From there, the scene shifts to older rocks formed in earlier oceans. Having established how we can study life's deeper history, we explore the fragmentary record of Earth's earliest organisms and ruminate on the origins of life, before ascending again through geological time, following a trail of fossils and molecules that leads back to the Cambrian "Explosion" of animal life, now seen as both the culmination of life's long Precambrian history and a radical departure from it.

I have three goals in writing this book. First is the obvious one. "Narrative history," wrote C. Vann Woodward, is "the end product of what historians do. The narrative is where they put it together and make sense for the reader." To me, science's creation story is a deeply engrossing narrative that, told correctly, helps us to understand not only our biological past, but the Earth and life that surround us today. Contemporary biological diversity is the product of nearly 4 billion years of evolution. We are a part of this legacy. Thus, by coming to grips with life's long evolutionary history, we begin to understand something of our own place in the world, including our responsibility as planetary stewards.

My second goal is to tell the story of early evolution in a particular way. The history of life is commonly recounted as a naturalist's Generations of Abraham: bacteria begat protozoans, protozoans begat invertebrates, invertebrates begat fishes, and the like. Such catalogs of received wisdom can be memorized, but there isn't a lot to think about. For this reason, I have chosen to relate my story as an enterprise—one in which rocks and fossils are encountered in remote corners of the globe, analyzed in the laboratory, and interpreted in light of processes (but not necessarily conditions) observable today. Discoveries in paleontology, the most traditional of scientific pursuits, weave together with emerging insights from molecular biology and geochemistry.

In some ways, conventional paleontology and the research described here seem poles apart, their practitioners squinting to view the past through the opposite ends of a telescope. Dinosaur bones are big and spectacular—they keep you *awake* at night. But, apart from the size of its inhabitants, the *world* of the dinosaurs was much like our own. In contrast, Earth's deep history is recounted by microscopic fossils and subtle chemical signals. And yet, the story they tell is dramatic, a succession of vanished worlds that leads through atmospheric transformation and biological revolution to the Earth we know today.

If we want to understand events that took place a billion or more years ago, how do we go about doing it? It's one thing to learn that photosynthetic bacteria lived on tidal flats 1.5 billion years ago, and quite another to understand how we recognize microscopic fossils as photosynthetic, how we determine that the rocks enclosing them formed on an ancient tidal flat, and how we estimate their age as 1.5 billion years. The epistemological leitmotif of how we know what we think we know recurs throughout this book. As human enterprise, this is also a story of exploration that extends from the inner space of molecules to the literal outer space of Mars and beyond. Cold nights in Siberia are part of the tale, as are warm friendships in China.

Finally, having excavated and evaluated the preserved shards of our biological past, I want to step back and ask whether we can identify any general principles that shine through the maze of historical particulars. What are the grand themes of life's early history? The astrobiologist in me, eager for a glimpse of samples collected on Mars, asks what aspects of our terrestrial biology might be found wherever life exists and which features are likely to prove the specific products of our particular planetary history? We don't yet know the answer, but how we search for life elsewhere in the universe depends to a large extent on how we think about this question.

One clear theme of evolutionary history is the cumulative nature of biological diversity. Individual species (of nucleated organisms at least) may come and go in geological succession, their extinctions emphasizing the fragility of populations in a world of competition and environmental change. But the history of guilds—of fundamentally distinct morphological and physiological ways of making a biological living—is one of accrual. The long view of evolution is unmistakably one of accumulation through time, governed by rules of ecosystem function. The replacement series implied by the Generations of Abraham approach fails to capture this basic attribute of biological history.

Another great theme is the coevolution of Earth and life. Both organisms and environments have changed dramatically through time, and more often than not they have changed in concert. Shifts in climate, in geography, and even in the composition of the atmosphere and oceans have influenced the course of evolution, and biological innovations have, in turn, affected environmental history. Indeed, the overall picture that emerges from our planet's long history is one of *interaction* between organisms and environments. The evolutionary epic recorded by fossils reflects, as much as anything else, the continuing interplay between genetic possibility and ecological opportunity.

This long view of biological history provides what may be the grandest theme of all. Life was born of physical processes at play on the young Earth. These same processes—tectonic, oceanographic, and atmospheric—sustained life through time as they shaped and reshaped our planet's surface. And, eventually, life expanded and diversified to become a planetary force in its own right, joining tectonics and physical chemistry in the transformation of air and oceans. To me, the emergence of life as a defining—perhaps *the* defining—feature of our planet is extraordinary. How often has this happened in the vastness of the universe? That's what *I* think about when I look up "in perfect silence at the stars."

Awe and humility attended the telling of earlier creation stories. They are appropriate companions to science's version, as well.

1 | In the Beginning?

Fossils found along the Kotuikan River in northern Siberia document the Cambrian "Explosion," the remarkable flowering of animal life that began some 543 million years ago. As Charles Darwin recognized more than a century ago, Cambrian fossils raise fundamental questions about life's earlier evolution. What kind of organisms preceded these already complex animals? Can we find older rocks, and if we can, will they preserve a record of Earth's earliest biological history?

Sometimes the past was shot with a hand-held camera; sometimes it reared monumentally inside a proscenium arch with moulded plaster swags and floppy curtains; sometimes it eased along, a love story from the silent era, pleasing, out of focus and wholly implausible. And sometimes there was only a succession of stills to be borrowed from the memory.
—Julian Barnes
Staring at the Sun

THE CLIFFS ALONG the Kotuikan River glow fawn and pink in the late afternoon sun (figure 1.1). Elsewhere, in North America or in Europe, a vista like this would be celebrated as a national park, its approaches flanked by campgrounds and souvenir shops. But here, in the forested wilderness of northern Siberia, its pastel beauty is both unremarkable and largely unseen. From a sheltered niche halfway up the cliff, I look up at my friend Misha Semikhatov perched high above the river, his large frame barely supported by a narrow ledge. The drop beneath his

Figure 1.1. Fossiliferous cliffs along the Kotuikan River in Siberia. The distance from river level to the top of the cliff is more than 300 feet, recording some 20 million years of Early Cambrian history.

feet is precipitous, but Misha's attention is elsewhere, fixed on a layer of sedimentary rocks just above his head. To his experienced eye, the bed of crinkly laminated limestones tells of an ancient tidal flat that bordered a vanished ocean, a broad expanse of shoreline exposed at low tide, covered by thickly matted bacteria, and occasionally crossed by small animals. As I rest against the rock face, observing marginally older beds, jotting in my notebook, and swatting mosquitoes (not necessarily in that order), I reflect on what has brought Misha and me to this remote spot high above the Arctic Circle (figure 1.2). The literal answer is a giant Soviet-era military helicopter that deposited us, a small group of colleagues, and a ton of gear some seventy miles upstream. From there, small rubber rafts floated us like Huckleberry Finn slowly down the river, through canyons of limestone, beneath circling falcons, past wolves that howl at the midnight sun, to this wild and beautiful place.

Of course, helicopters provide only one of several appropriate responses to the question of what brought us here. The deeper and more interesting answer is that these cliffs, cut over the millennia by the Kotuikan as it winds toward the Arctic Ocean, record one of Earth history's great turning points. As well as any rocks known anywhere, they document the remarkable diversification of animal life popularly known as the Cambrian Explosion. In the broadest possible sense, the Kotuikan cliffs record the beginnings of the modern world, a world in which animals swim, crawl, or walk beneath an atmosphere of breathable air. That's really what brought us here.

At river level, a series of steps rise out of the water like prehistoric ghats, hewn by nature from thin beds of limestone and dolomite. Some

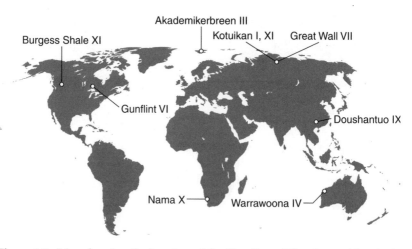

Figure 1.2. Map showing the location of the Kotuikan cliffs, along with principal localities discussed in subsequent chapters (denoted by roman numerals).

545 million years ago, these rocks were deposited as lime muds in a warm shallow seaway not unlike the modern Florida Keys. Scattered clusters of gypsum crystals record drying that episodically left coastal waters salty enough to exclude all but the hardiest bacteria. The fossils of animals are rare in these rocks, and those that can be found are simple. Only a few irregular meanders disturb bedding surfaces, the trails of small wormlike creatures that crawled along the muddy bottom in search of food.

About ten feet above the river, an abrupt shift to quartz sandstone marks the so-called Precambrian-Cambrian[1] boundary, historically the line of demarcation between the tractable paleontology of the Phanerozoic Eon (literally, the "age of visible life") and the terra incognita of a more youthful Earth. Volcanic rocks a few hundred miles to the east date this horizon at 543 (plus or minus one) million years before the present. Above the sandstone bench purple, red, and green shales form a steep shoulder above which vertiginous cliffs of limestone rise like a wall. The shales record a flooding event, with shoreline sands pushed far to the west by the rising sea. As sediments accumulated, the sea again grew shallower, so that the overlying limestone beds record environments progressively closer to the ancient shore. Near the top of the cliff face, an irregular surface marks a point at which the sediments were exposed and eroded by some vanished forebear of the Kotuikan River, only to be drowned again as the sea reclaimed lost territory.

Beginning at the level of the sandstone bench, the rocks contain small skeletal fossils. In the lowermost beds, there are only a few forms, hollow cones of calcite little more than a millimeter long (figure 1.3a). But as we ascend the cliff, slowly and carefully to avoid a career-shortening slip, the abundance and variety of these fossils increase. So, too, do the number and behavioral complexity of preserved tracks, trails, and burrows. Near the top of the cliff, more than three hundred

[1] By convention, geologic time is divided into four eons: the Phanerozoic (0–543 million years ago), the Proterozoic (543–million 2.5 billion years ago), the Archean (2.5–ca. 4 billion years ago), and the Hadean (the time interval from the accretion of the Earth to the beginning of the preserved record, ca. 4–4.55 billion years ago). The Cambrian is the initial period of the Phanerozoic Eon; thus, all earlier time is commonly, if informally, referred to as "Precambrian." See geologic timescale on page 2.

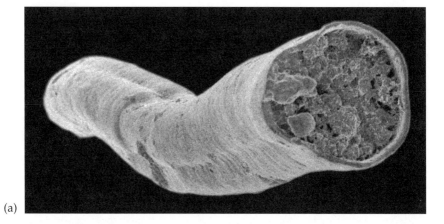

(a)

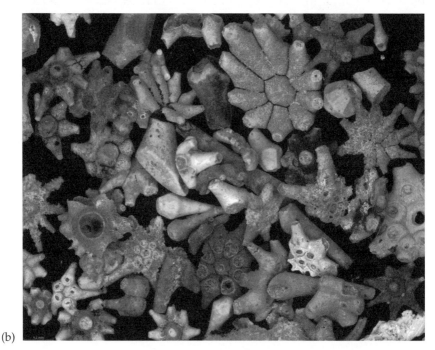

(b)

Figure 1.3. Small shelly fossils in basal Cambrian rocks. (a) *Anabarites trisulcatus*, the tiny skeletons found in lowermost Cambrian beds along the Kotuikan River. These specimens come from rocks of comparable age in China. (b) Small shelly fossils of the types found higher in the Kotuikan cliffs; most of the forms seen here are the skeletal spicules of chancellorids, enigmatic baglike animals found widely (and only) in Cambrian rocks. (Images courtesy of Stefan Bengtson)

feet above river level, rocks estimated to be about 525 million years old contain nearly one hundred different types of shells (figure 1.3b). Some, like the small cones in the cliff base, have a threefold symmetry that differentiates them from most animals alive today. Others, however, include small spiral shells that are recognizably the remains of mollusks, bivalved skeletons formed by brachiopods, and, a little higher up, the segmented bodies of trilobites. Painstakingly collected and described by Russian paleontologists, these fossils chronicle an apparently rapid unfolding of biological diversity in the Cambrian ocean. In less than 20 million years, the seafloor was transfigured from the alien to the (at least broadly) familiar. The same drama is recorded in rocks of comparable age throughout the world, providing our earliest glimpses of the animals that have populated Earth's oceans ever since that time.

Charles Darwin couldn't get this pattern out of his mind. One might suppose that Darwin, like his modern intellectual descendants, saw in the fossil record a confirmation of his theory—the literal documentation of life's evolution from the Cambrian to the present day. In fact, the two chapters devoted to geology in *The Origin of Species* are anything but celebratory. On the contrary, they constitute a carefully worded apology in which Darwin argues that evolution by natural selection is correct *despite* an evident lack of support from fossils.

Darwin envisioned natural selection as a slow but continuous process by which biological lineages diverged and gradually grew more distinct from one another. Intermediate forms that link different species are rare in the modern world because selection inexorably acts against them. But why don't we see intermediates in time? Darwin's expectation was that successive sedimentary beds should document the gradual transition from one form, perhaps seen at the base of a cliff, to its morphologically distinct descendants at the top. That such series are rare he attributed to the extreme imperfection of the fossil record.

The *Origin* is full of magisterial prose, words that are luminous as well as illuminating. Darwin's characterization of the geological record is particularly striking: "a history of the world imperfectly kept, and written in a changing dialect; of this history we possess the last volume alone, relating only to two or three countries. Of this volume, only here and there a short chapter has been preserved; and of each page only here and there a few lines."

Darwin might well have embraced Julian Barnes's description of human remembrance as a metaphor for Earth's geological memory: sedimentary rocks provide a succession of widely spaced snapshots, not a documentary film of our planetary history. At the local level observed in a roadside or cliff face, this view is well justified, and Darwin's arguments seem strikingly modern. We understand today that sedimentary rocks provide discontinuous records in which the boundary between two layers may represent more time than the beds themselves. But the geometry of sedimentary accumulation is more complex than the orderly layer cake commonly seen in local outcrop. Viewed three-dimensionally, the layers pinch and swell like hills in a van Gogh landscape, thickening here and changing character, thinning there to feather edge. Time represented in one place by a hiatus between beds is recorded elsewhere by sediment accumulation. Viewed still more broadly, at any point in time such locally discontinuous records are forming in many basins throughout the world. Thus, if we revisit Darwin's metaphor for geological history, we find that while his book is missing chapters, and the chapters in hand are missing pages, we actually possess multiple copies of the text and the parts that are missing vary from copy to copy. If we have a principle for interleaving the surviving pages, it is possible to stitch together a composite record that, for the past 600 million years, at least, isn't bad. The discipline of stratigraphy provides that principle, showing us that at least the broad biological patterns read from fossils reflect evolution and not gross inadequacies of the rock record.

Biostratigraphers have known for more than a century that species commonly appear in the fossil record fully formed, persist without much change for million of years, and then disappear. The sense conveyed by this pattern, that form changes episodically rather than continuously, doesn't arise because species appear only once, in a single bed. It is justified because species commonly occur in many successive beds with little change, or at least little *directional* change, from bottom to top—a pattern we can't explain away as the product of sedimentary incompleteness (at least not without making assumptions that, in many cases, we know to be implausible). Recognizing this, Niles Eldredge and Stephen Jay Gould argued in 1972 that it is this stratigraphic pattern of "punctuated equilibrium"—and not Darwin's picture of gradual change—that is most consistent with modern evolutionary theory. Most new species arise not from the insensibly gradual transformation of large populations but rather by the rapid differentiation of small, isolated populations at the periphery of the main group. The transformations envisioned by Darwin occur, but they take place rapidly and locally, after which populations of the descendant species are constrained by natural selection to stay more or less the same until competitors or shifting environments spell their doom.

The everyday comings and goings of fossil species can be reconciled with both evolutionary expectation and geological reality, but what about the spectacular pattern seen in the Kotuikan cliffs? How do we explain this biological transformation of the oceans? If Darwin was concerned about the general lack of transitional forms in the fossil record, he was truly disquieted by the apparently abrupt appearance of abundant, diverse, and anatomically complex animals in the oldest Cambrian beds:

> There is another and allied difficulty which is much graver. I allude to the manner in which numbers of species of the same group, suddenly appear in the lowest known fossiliferous rocks. . . . The case must at present remain inexplicable; and may be truly urged as a valid argument against the views herein entertained.

Of course, the *Origin* does offer an explanation, and it is the one we might expect—massive record failure at the base of the Cambrian System. It isn't, wrote Darwin, that no life preceded the fabulously complicated snails and trilobites of the Cambrian, but rather that their ancestors' record lay in older beds that are deeply buried, destroyed, or undiscovered. In another memorable passage, Darwin insisted that

> if my theory be true, it is indisputable that before the lowest Silurian[2] stratum was deposited, long periods elapsed, as long as, or probably far longer

[2] In the mid–nineteenth century, debate about how to define and differentiate the Cambrian and Silurian Systems remained unresolved. Darwin employed Londoner Roderick Murchison's term Silurian for the oldest fossiliferous beds, despite the fact that his Cambridge mentor Adam Sedgwick had coined the name "Cambrian." Not until 1879 did Charles Lapworth cut the Gordian knot, retaining Cambrian for the lower part of the disputed system, Silurian for its upper portion, and Ordovician (after the Ordovici, an ancient and putatively obstreperous Welsh tribe) for the contested interval of overlap.

Figure 1.4. Upstream along the Kotuikan River, showing the angular unconformity between the latest Precambrian-Cambrian succession seen in figure 1.1 and an older package of sedimentary rocks that lies beneath it.

> than, the whole interval from the Silurian age to the present day; and that during these vast, yet quite unknown periods of time, the world swarmed with living creatures.

Back along the Kotuikan River, Misha and I sit on a gravel bar opposite the cliffs and consider Darwin's dilemma as we sip our evening tea. How could such complexity evolve so quickly? And if it didn't really happen so fast, where are the rocks that record life's earlier history?

The sedimentary beds in the Kotuikan cliffs aren't quite flat-lying; tectonic movements over millions of years have tilted them slightly downward to the west. Because of this, a hike toward the east, upstream along the river, reveals layers that sit ever farther below the level of the Cambrian fossils. About fifteen miles up river—some 200 feet lower in the sedimentary rock column—we encounter a sharp stratigraphic break, the base of the sedimentary package that includes the latest Precambrian carbonate rocks and basal Cambrian animals (figure 1.4). Is that the end of the sedimentary trail?

Not at all. What lies beneath these rocks is another, older succession of sandstones, shales and carbonates. Set at an acute angle to the younger beds, this older package is itself more than 3,500 feet thick. The base of the Cambrian System is not the bottom of the stratigraphic record—not in northern Siberia, and not in many other places where tectonic circumstance has preserved sedimentary rocks deposited one, two, or even 3 billion years before Cambrian beds began to accumulate.

We can put Darwin's conjecture to the test. Is the Cambrian Explosion the beginning of biological history? Or is it the culmination of evolutionary events that extend much deeper into our planet's past?

2 | The Tree of Life

In the Tree of Life, built from comparisons of nucleotide sequence in genes from diverse organisms, plants and animals form only small twigs near the top of one branch. Life's greater diversity, and, by implication, its deeper history, is microbial. If we wish to explore Precambrian rocks for evidence of early life, we must first learn about Bacteria and Archaea, the tiny architects of terrestrial ecosystems.

Most of us learn about Richard III through Shakespeare's eponymous drama, but as history, this account is suspect—after all, Shakespeare's patrons *won* the War of the Roses. Biased, selective, incomplete, and even incomprehensible documents are the daily bread of historians. Despite the shortcomings of individual accounts, however, scholars can arrive at a balanced understanding of the past by sifting through a number of different records for points of agreement and complementary perspectives.

The study of biological history works much the same way. The fossiliferous cliffs along the Kotuikan River served to introduce one great library of Earth's evolutionary past—the geological record. Sedimentary rocks preserve a remarkable record of life and environments through time, but as we've already observed, this accounting is episodic, not continuous. It is also highly selective, brightly illuminating some groups of organisms while leaving others in darkness. For example, we know a great deal about the paleontology of horses, but little about the earthworms beneath their feet.

Fortunately, we can consult a second library—the biological diversity that surrounds us today. Comparative biology offers rich resources for evolutionary analysis, providing genealogy to complement paleontology's record of time, and physiology to match geology's chronicle of environmental change. The great cell biologist Christian de Duve has gone so far as to suggest that the genes of living organisms contain a *full* accounting of evolutionary history. If so, however, it is—like Shakespeare's histories—limited to an account of life's winners. Only paleontology can tell us about trilobites, dinosaurs, and other biological wonders that no longer grace the Earth. If we wish to understand life's history, then, we must weave together insights drawn from geology *and* comparative biology, using living organisms to reanimate fossils and fossils to learn how the diversity of our own moment came to be.

Despite an almost bewildering diversity of form and function, all cells share a common core of molecular features, including ATP (life's principal energy currency), DNA, RNA, a common (with a few minor exceptions) genetic code, molecular machinery for transcribing genetic information from DNA into RNA, and more machinery to translate RNA messages into proteins that provide structure and regulate cell function. The reciprocal observation is equally striking. In spite of their fundamental unity of molecular structure, organisms display extraordinary variation in size, shape, physiology, and behavior. Life's unity and diversity are both remarkable in their own ways; together they comprise the two great themes of comparative biology.

Even a casual observer will notice the pattern of nested similarity displayed by Earth's biological diversity. Humans and chimpanzees are clearly distinct, but they share many features of anatomy and physiology, resembling each other far more than either does, say, a horse. Humans, chimps, and horses, in turn, share features such as hair, lungs, and limbs that separate them from catfish. Yet, all animals with bony skeletons share a basic pattern of anatomical organization that unites them as a group and differentiates them from other sets of species built on different design principles—insects, for example, or spiders.

The nested similarity of species was well known to early naturalists. Linnaeus codified it in the 1730s, proposing a hierarchical system of taxonomic classification that is still in use today. It was Charles Darwin, however, who explicitly recognized the genealogical nature of this pattern. Biological differences have arisen through time, he wrote, because of "descent, with modification," that is, by evolutionary divergence from common ancestors under the influence of natural selection:

> The affinities of all the beings of the same class have sometimes been represented by a great tree. I believe this simile largely speaks the truth. The green and budding twigs may represent existing species; and those produced during each former year may represent the long succession of extinct species. . . . As buds give rise by growth to fresh buds, and these, if vigorous, branch out and overtop on all sides many a feebler branch, so by generation I believe it has been with the great Tree of Life, which fills with its dead and broken branches the crust of the earth, and covers the surface with its ever branching and beautiful ramifications.

We can explain the similarities between humans and chimps by descent from a common ancestor that possessed the various features the two groups share. Their differences have arisen since they diverged. This makes the paleontological prediction that the oldest fossils of humanlike primates should resemble the last common ancestor of chimps and humans more closely than modern people do; the features that make us distinctly human should appear only in younger fossils of our lineage. The fossil record of human ancestry is notoriously sketchy, but skeletal remains unearthed in Africa and Asia confirm this prediction. (Note that there is no expectation that successively older members of our lineage should close in on chimp morphology. Humans didn't descend from chimpanzees; humans and chimps *both* diverged from a common ancestor that was neither *Homo* nor *Pan*.)

Not all shared features are equally helpful in determining "propinquity of descent" (another delightful Darwinism). For example, birds, bats, and the extinct pterosaurs all sport wings, but their wings have different skeletal structures, and many other features show that these airborne animals are not closely related. Wings evolved independently in each group as an adaptation for flight; in the parlance of systematic biology, these features are *convergent*. Only features that are shared because of common ancestry (*homologies*, in evolution-speak) can be used to assess evolutionary relationships. In practice, we don't always know whether similar features are convergent or homologous and so rely on sophisticated computer algorithms to sort out large sets of comparative biological data.

It is relatively easy to see how morphological characteristics might be used to articulate a hypothesis of evolutionary relatedness, or *phylogeny*, for all primates, all mammals, or even all vertebrate animals. We can also grant that an expert, at least, could do the same for mollusks or arthropods. But how can we place mollusks, arthropods, *and* vertebrates within a greater evolutionary tree of all animals? And, much harder, how can we reconstruct the whole of Darwin's great Tree of Life, a phylogeny that encompasses all living things?

Wandering through an alpine forest or snorkeling above a coral reef, we observe an ecology shaped by plants (or seaweeds) and animals, with large vertebrates at the top of the food chain and other creatures below. Ecosystems also contain many organisms that we can't see, but concern for their contributions is generally fleeting—surely bacteria and other microorganisms, tiny and simple, eke out their living in a world of our making?

As large animals, we can be forgiven for holding a worldview that celebrates ourselves, but, in truth, this outlook is dead wrong. We have evolved to fit into a bacterial world, and not the reverse. Why this should be is, in part, a question of history, but it is also an issue of diversity and ecosystem function. Animals may be evolution's icing, but bacteria are the cake.

Plants, animals, fungi, algae, and protozoa are *eukaryotic* organisms, genealogically linked by a pattern of cell organization in which genetic material occurs within a membrane-bounded structure called the nucleus. Bacteria and other *prokaryotes* are different—their cells lack nuclei. In terms of biological importance, eukaryotes would seem to have a decisive edge; eukaryotic organisms display a variety of form that ranges from scorpions, elephants, and toadstools to dandelions, kelps, and amoebas. In contrast, prokaryotes are mostly minute spheres, rods, or corkscrews. Some bacteria form simple filaments of cells joined end to end, but very few are able to build more complicated multicellular structures.

Size and shape surely favor eukaryotes, but morphology provides only one of several yardsticks for measuring ecological significance. Metabolism—how an organism obtains materials and energy—is another, and by this criterion, it is the prokaryotes that dazzle with their diversity. Eukaryotic organisms basically make a living in one of three ways. Organisms like ourselves are *heterotrophs*; we gain both the carbon and energy needed for growth by ingesting organic molecules made by other organisms. To obtain energy, our cells use oxygen to break down sugars to carbon dioxide and water, a process called *aerobic* (oxygen-using) *respiration*. In a pinch, we can gain a bit of energy from a second metabolism called *fermentation*, an *anaerobic* (no-oxygen) process in which one organic molecule is broken into two others—only brewer's yeast and a few other eukaryotes make much of a living this way. The third principal energy metabolism found in eukaryotes is the *photosynthesis* performed by plants and algae: chlorophyll and associated pigments harvest energy from sunlight, enabling plants to fix carbon dioxide into organic matter. In order to convert light into biochemical energy, plants need an electron—water supplies the needed charge, producing oxygen as a byproduct.

A *Christmas Carol*, Charles Dickens's classic tale of redemption, opens with an admonition for readers to pay close attention to a particular fact: "Old Marley was as dead as a door-nail. . . . This must be distinctly understood, or nothing wonderful can come of the story I am going to relate." The early history of life has its own "Jacob Marley" facts that, like the old miser's death in Dickens's story, need to be understood if the narrative is to make sense. First of these is the metabolic diversity of prokaryotic microorganisms, key to any exploration of early biological history. We must come to grips with the many ways that prokaryotes make a living and how these tiny organisms fit onto the Tree of Life before we put our boots back on and return to the field as paleontologists.

Like eukaryotes, many bacteria respire using oxygen. But other bacteria can respire using dissolved nitrate (NO_3^-) instead, and still others use sulfate (SO_4^{2-}) ions or the metallic oxides of iron and manganese. A few prokaryotes can even use CO_2 to react with acetic acid, generating natural gas, which is methane (CH_4). Prokaryotic organisms have also evolved a galaxy of fermentation reactions.

Bacteria ring changes on the theme of photosynthesis, as well. The cyanobacteria, a group of photosynthetic bacteria tinted blue-green by chlorophyll and other pigments, harvest sunlight and fix CO_2 much like eukaryotic algae and land plants. However, when hydrogen sulfide

(H_2S, well known for its "rotten egg" smell) is present, many cyanobacteria use this gas rather than water to supply the electrons needed for photosynthesis. Sulfur and sulfate are formed as by-products, but oxygen is not.

The cyanobacteria comprise only one of five distinct groups of photosynthetic bacteria. In the other groups, electron supply by H_2S, hydrogen gas (H_2), or organic molecules is obligatory, and oxygen is never produced. These photosynthetic bacteria harvest light using bacteriochlorophyll rather than the more familiar chlorophyll. Some employ the same biochemistry as cyanobacteria and green plants to fix carbon dioxide, but others have distinctively different pathways, and still others rely on carbon already packaged into organic molecules.

Bacterial variations on the metabolic themes of respiration, fermentation, and photosynthesis are, thus, impressive, but prokaryotic organisms have evolved yet another way of growing that is completely unknown in eukaryotes: *chemosynthesis*. Like photosynthetic organisms, chemosynthetic microbes get their carbon from CO_2, but they harvest energy from chemical reactions rather than sunlight. Oxygen or nitrate (or, less commonly, sulfate, oxidized iron, or manganese) is combined with hydrogen gas, methane, or reduced forms of iron, sulfur, and nitrogen in ways that allow the cell to capture the energy released by the reaction. Methanogenic prokaryotes are of particular evolutionary and ecological interest; these tiny cells can gain energy from the reaction of hydrogen gas and carbon dioxide to produce methane.

The metabolic pathways of prokaryotes sustain the chemical cycles that maintain Earth as a habitable planet. Take carbon dioxide, for example. Volcanoes supply CO_2 to the oceans and atmosphere, but photosynthesis removes it at a far faster clip. So much faster, in fact, that photosynthetic organisms could strip the present-day atmosphere of its CO_2 in little more than a decade. They don't, of course, principally because respiration, in essence, runs the photosynthetic reaction backward. While photosynthetic organisms react CO_2 and water to produce sugar and oxygen, respiring creatures (including you, as you read this sentence) react sugar with oxygen, giving off water and carbon dioxide. Together, photosynthesis and respiration *cycle* carbon through the biosphere, sustaining life and maintaining the environment through time.

It is easy to envision a simple carbon cycle in which cyanobacteria fix CO_2 into organic matter and supply oxygen to the environment, while respiring bacteria do the reverse, consuming oxygen and regenerating CO_2. Plants and algae would do just as well as cyanobacteria, and protozoa, fungi, and animals could substitute for bacterial respirers—the prokaryotes and eukaryotes are functionally equivalent. But let's let some cells sink to the seafloor and become buried in oxygen-depleted sediments. Now the limitations of eukaryotic metabolism become clear—reactions that do not use oxygen (*anaerobic* reactions) are needed to complete the carbon cycle. In modern seafloor sediments, sulfate reduction and respiration using iron and manganese are just as important as aerobic respiration in recycling organic matter. More generally, wherever carbon passes through oxygen-free environments, bacteria are essential to the carbon cycle; eukaryotes are everywhere optional.

The fundamental importance of prokaryotes extends to other biologically important elements, as well. Indeed, in the biogeochemical cycles of sulfur and nitrogen, *all* the principal metabolic pathways that cycle these elements are prokaryotic. Consider, in particular, nitrogen, an essential element required for the formation of proteins, nucleic acids, and other biological compounds. We live our lives bathed in nitrogen gas. (Air is about 80 percent N_2 by volume.) But this vast repository of nitrogen is not biologically available to us; like other animals, we obtain the nitrogen we need by eating other organisms. As it turns out, nitrogen gas is no more available to cattle or corn than it is to humans. Plants can take up ammonium (NH_4^+) or nitrate from the soil, but how do these compounds get there in the first place? Ammonium is released as dead cells decay; nitrate, in turn, is produced by bacteria that oxidize ammonium. In oxygen-rich habitats, the resulting nitrate is available to plants (or, in aquatic ecosystems, algae and cyanobacteria), but in waterlogged soil or other environments where O_2 is depleted, other bacteria use nitrate for respiration, returning nitrogen to the atmospheric pool of N_2. (Much of the nitrate spread across fields as fertilizer is lost in this way.)

So, we haven't solved our problem. The ammonium and nitrate in soil and seawater come from dead cells, and nitrate-respiring bacteria inexorably remove biologically usable nitrogen from the environment. What, then, fuels the biological nitrogen cycle and keeps it from running down?

The answer is that some organisms are able to convert atmospheric nitrogen to ammonium, using the cell's store of energy. *No* eukaryotic organism can fix nitrogen in this way, but many prokaryotes can. (Farmers commonly include soy or other beans in their crop rotation because these plants restore nitrogen to the soil. The task of nitrogen fixation, however, is accomplished by bacteria that live in small nodules on the roots of bean plants, not by the beans themselves.) A small amount of nitrogen is fixed by lightning as it cuts through the atmosphere, but biology's thirst for nitrogen is quenched mainly by bacteria.

The cycles of carbon, nitrogen, sulfur, and other elements are linked together into a complex system that controls the biological pulse of the planet. Because organisms need nitrogen for proteins and other molecules, there could be no carbon cycle without nitrogen fixation. Nitrogen metabolism itself depends on enzymes that contain iron; thus, without biologically available iron, there could be no nitrogen cycle . . . and, hence, no carbon cycle. Biology on another planet may or may not include organisms that are large or intelligent, but wherever it persists for long periods of time, life will feature complementary metabolisms that cycle biologically important elements through the biosphere.

By now it should be apparent why I insisted earlier that plants and animals evolved to fit into a prokaryotic world rather than the reverse. It *is* a prokaryotic world, and not only in the trivial sense that there are a lot of bacterial cells. Prokaryotic metabolisms form the fundamental ecological circuitry of life. Bacteria, not mammals, underpin the efficient and long-term functioning of the biosphere.

How can this astonishing diversity of prokaryotic cells be ordered and assembled along with that of eukaryotes into a phylogeny that encompasses all of biology? Size and shape fail us, and so does physiology; organisms as disparate as fungi and elephants, or *E. coli* and redwoods, are simply too different from one another to assemble into a believable tree based on form and function alone. The solution requires that we return to the unity of life, the molecular attributes shared by all known organisms. In a groundbreaking paper published in 1965, Emile Zuckerkandl and Nobel laureate Linus Pauling proposed that molecules can be read as documents of evolutionary history. Just as the anatomical structures of limbs or skulls reflect descent with modification, so too do the chemical structures of DNA and proteins. The long chain of amino acids that makes up, say, the respiratory protein cytochrome c differs slightly between humans and chimps and more so between humans/ chimps and horses. The sequences of nucleotides in the genes that code for these proteins differ correspondingly.

Carl Woese of the University of Illinois built decisively on this conceptual foundation. Woese spent his formative years in science investigating ribosomes, the sites within cells where proteins are manufactured. He knew that all organisms contain ribosomes, that all ribosomes contain functional complexes made of RNA and proteins, and that these complexes all contain several subunits. By comparing among organisms the sequences of nucleotides that make up the RNA molecules found in the small subunit of ribosomes, Woese made the great leap that brought phylogeny to the microbial world, sowing the seeds for a Tree of Life worthy of the name.

Figure 2.1 shows the Tree of Life, a depiction of the genealogical relationships of all living organisms, based on comparisons of molecular sequence in the genes that code for small subunit ribosomal RNA. Experts argue about its details, but all biologists agree that our ability to draw Darwin's great Tree of Life in its entirety constitutes one of the great intellectual achievements of the late twentieth century.

The first thing to notice about the tree is that it contains three major limbs, termed *domains* by Woese. Two of the domains are unsurprising: the eukaryotes and the bacteria fall on distinct branches. The third, however, came as a shock when Woese and then postdoctoral fellow George Fox proposed it in 1977. The Archaea are prokaryotic in cell organization, and for many years the organisms on this branch had been thought of (when they were thought about at all) as metabolically unusual bacteria. But comparison of ribosomal RNA genes suggests that these microbes are fully as distinct from the conventional bacteria as bacteria are from eukaryotes. What's more, the tree indicates that archaeans are actually more closely related to the eukaryotes than they are to bacteria. (In phylogenetic discourse, closeness of relationship reflects recency of common ancestry; it is a statement about genealogy, not similarity.)

The complete genome (genetic information encoded in DNA) of the archaean *Methanococcus janaschii* was published in 1996, revealing that

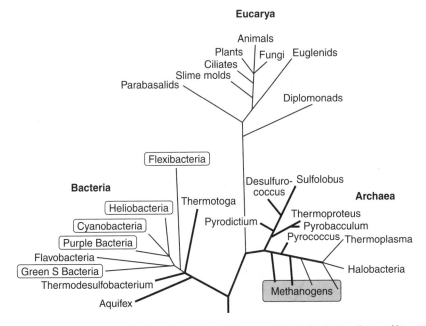

Figure 2.1. The Tree of Life, a depiction of the genealogical relationships of living organisms, based on sequence comparisons of genes that code for RNA in the small subunit of the ribosomes found in all cells. Note the three principal branches, made up of Bacteria, Archaea, and Eucarya. Branch lengths indicate degree of difference among gene sequences; because genes can evolve at different rates, however, this does not necessarily translate into time. Bacterial groups with photosynthetic members are highlighted by clear boxes; methanogenic archeans fall within the shaded box. Heavy lines denote hyperthermophiles—groups of organisms that live at high temperatures. (Adapted from a depiction of the tree by Karl Stetter)

this microbe shares just 11–17 percent of its genes with bacteria whose genomes have been sequenced. More than 50 percent of its genes are unknown in *either* eukaryotes or bacteria, confirming that archaeans are distinctly different from organisms in the other two domains. Archaeans do, however, have some important characters in common with bacteria, such as (most obviously) prokaryotic cell organization, the molecular structure of the ribosome, and the arrangement of genes on a single circular chromosome. Equally, on the other hand, archaeans share attributes such as molecular details of DNA transcription and susceptibility to specific antibiotics with eukaryotes. And there are still other traits that bacteria and eukaryotes share to the exclusion of Archaea—prominent among these is the nature of the cell membrane.

How then do we tell who is more closely related to whom? Put another way, where do we place the *root* on this tree? A three-branched tree can't be rooted by conventional means, and a bit more consideration of character distributions shows why. Features such as ATP or the genetic code that are shared by all three domains carry no information on genealogical relationships, but permit inferences about the nature of the last common ancestor of the three branches. In contrast, attributes such as cell wall composition that are distinct in each limb provide no information on either genealogy or ancestral features. Characters shared by two of the three domains would appear to offer better prospects for tree building, but such distributions can be explained equally well in several different ways. For example, if we assume that membranes composed of fatty acids were present in the last common ancestor, then we can posit that this trait was retained in bacteria and eukaryotes but replaced by isoprenoid-based membranes along the road to the Archaea. Alternatively, we can assume that membranes built from isoprenoids are ancestral, but were swapped for fatty-acid membranes in the common ancestor of bacteria and eukaryotes. Like the first alternative, this tree requires only one evolutionary change. (We could, of course, eliminate some possibilities if we knew for certain which traits characterized the last common ancestor, but we have no way of determining this.)

A clever solution to the rooting problem was proposed in 1989. Two research groups headed by Naoyuki Iwabe and Peter Gogarten, respectively, independently recognized that while three sets of organisms cannot be joined into a rooted tree, some of the genes they contain can be. The genes in question share a specific property: they were present in duplicate sets in the last common ancestor. How does this help us? As shown in figure 2.2, each of the two sister genes present in the last common ancestor diverged as the three domains differentiated. The resulting array of genes can be ordered into a tree. In toto, the tree is unrooted, but it consists of two component branches that can be rooted *relative* to each other. Both "half trees" have the same form: one branch that contains only bacteria and a second containing Archaea plus eukaryotes.

This exercise has been repeated numerous times, using dozens of gene families. Many trees yield the root position shown in figure 2.2, but

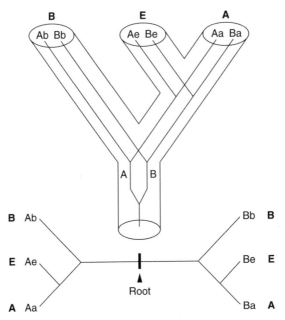

Figure 2.2. Rooting the Tree of Life. *Top:* The genealogical relationships among Bacteria (B), Eucarya (E), and Archaea (A) are shown by the hollow cylindrical branches. Lines within the cylinders show the phylogeny of a gene that duplicated into forms A and B prior to the differentiation of the three domains from their last common ancestor. *Bottom:* The evolutionary relationships among the genes are shown here. Each half tree can be rooted relative to the other, allowing molecular biologists to reconstruct the genealogical relationships among eukaryotes, archaeans, and bacteria.

others suggest different relationships among the three domains. No one tree satisfies all genetic data, forcing a startling conclusion. We think about genes being passed vertically through the tree from ancestor to descendant, but some genes must have been passed horizontally from one branch to another, perhaps by hopping a ride on a virus or by the uptake of DNA from dead cells. Contemporary organisms are, thus, genetic chimeras.

This revelation potentially casts doubt on the entire endeavor of constructing phylogenies from gene sequences, because gene trees and organism trees will not coincide when horizontal transfer has occurred. According to some biological Cassandras, the genes of microorganisms have been swapped so often and so promiscuously that no meaningful tree of microbial *organisms* can be recovered from molecular comparisons. This possibility is both frustrating and tantalizing, but it may be overdrawn. Sorel Fitz-Gibbons of UCLA and Christopher House, now at Penn State, analyzed the distribution of all genes in the dozen or so organisms whose complete genomes had been sequenced by early 1999. The tree recovered by comparing universal gene distributions closely matches that inferred from ribosomal RNA gene sequences, suggesting that despite gene swapping, phylogenetic order underpins the genomes of bacteria and archaeans.

James Lake and Maria Riviera, also of UCLA, have even hypothesized that certain rules govern the likelihood of horizontal transfer. *Informational* genes that code for basic features of cell biology appear to be unlikely candidates for lateral exchange—ribosomal RNA genes fall into this category. In contrast, *operational* genes, genes or groups of genes that encode specific metabolic functions, may be passed from one lineage to another with relative ease by means of viruses or other vectors. For example, we know that bacterial tolerance to heavy metals can be gained by the uptake of particular genes.

Dawn is just breaking on the study of microbial genes and phylogeny, and as more and more genomes are sequenced in their entirety, new insights can be expected to topple current generalizations. For now, it is reasonable to view the Tree of Life as a reflection of microbial genealogy. But we must think of it as a genealogy of the microbial "chassis," with specific features that "soup up" well-adapted organisms assembled, in part, by gene exchange across taxa.

The bacterial limb of the tree is profusely branched; at present we know of at least thirty major groups of bacteria, each more or less equivalent to the plant and animal kingdoms traditionally recognized by biologists. Most of the diverse metabolisms discussed in the previous section can be found on this limb. In particular, photosynthesis is a distinctly bacterial physiology. (How this squares with the obvious presence of photosynthesis in eukaryotic plants and algae is one of the great stories of evolutionary biology, but a tale best left for chapter 8.) Note, however, that photosynthetic lineages adorn only the upper branches of the bacterial limb. This suggests that Earth's earliest ecosystems were fundamentally different from those that surround us today. Today, photosynthesis fuels biology in most habitats. Early life, however, must have run on chemosynthesis. The earliest branches currently recognized on the bacterial limb contain chemosynthetic and heterotrophic organisms, many of which live at high temperatures in the absence or near absence of oxygen.

In contrast to Bacteria, the Archaea contain only two principal groups, albeit with hints of others yet to be characterized. One branch of archaeans is dominated by methanogenic organisms. Most bugs on this branch are obligate methane producers, but in at least three instances, individual lineages have evolved a more diverse metabolic repertoire that includes respiration. According to Gary Olsen of the University of Illinois, these "add-ons" are all encoded by genes transferred horizontally from bacteria. This emphasizes that horizontal transfer isn't something that occurred once or twice in the earliest history of life; it is a continuing and persistent means of creating biological novelty.

Closely related to the methanogenic archaeans are the halobacteria, a distinctive group of microorganisms that gain energy from the sun, using a light-harvesting pigment strikingly similar to the rhodopsin in vertebrate eyes; halobacteria obtain the carbon they need to grow by absorbing organic molecules. The other major branch of the Archaea includes organisms that derive energy from chemical reactions between hydrogen and sulfur compounds.

Archaea are widely distributed across the Earth, but we still know relatively little about most of them. For example, only in 2001 was it discovered that tiny archaeans may be the most abundant organisms in many parts of the ocean; biologists have no idea how these microbes make their living. On the other hand, some of the best-characterized Archaea live in unusual places—*very* unusual places. The halobacteria, for example, thrive in waters that are ten times saltier than the ocean. (The striking magenta sheen of halobacteria can be seen from the air in commercial salt ponds, such as those that line the landing approach to San Francisco airport.) Other archaeans live in acid mine waste with a pH of 1. And the current world record holder for temperature tolerance is *Pyrolobus fumarii*, an archaean that can grow in deep-sea hydrothermal vents at 113°C. (At the high pressures of the ocean bottom, water this hot remains liquid.) These *hyperthermophilic* organisms cannot grow at the temperatures used to pasteurize milk—not because they are too hot, but because they are too cold!

How do we think about such "extremophiles"? Are they merely fascinating oddballs, or do they tell us something fundamental about the history of life? Some, such as the halophiles, reside on distal branches of the Tree of Life, implying that these groups evolved relatively late in its history. In contrast, hyperthermophilic prokaryotes occupy a privileged position in the tree—they are found on the earliest branches of both the archaeal and bacterial limbs. This suggests that modern organisms are descended from ancestors that lived in hot environments. Thus, when we encounter microbial communities in the colorful hot springs of Yellowstone Park or in the midocean ridges that traverse the deep seafloor, we are glimpsing some of our earliest ancestors (color plate 1).

(Recently, geneticists have added an intriguing twist to this story. The amino acid sequences in proteins from living microorganisms can be used to reconstruct ancient proteins likely to have been present in the last common ancestor. Surprisingly, the putatively ancestral proteins synthesized from these reconstructions are not stable at high temperatures. If true, the last common ancestor of bacteria and archaeans could not have been hyperthermophilic at all. How to reconcile this finding with the Tree of Life remains a subject of debate. One possibility is that the earliest organisms evolved at moderate temperatures, but gave rise to (at least) two groups of descendants that colonized energy-rich hot springs. Now all we need is an exterminating angel to wipe out all life save the handful of lineages sheltered in hydrothermal foxholes. Giant meteor impacts would do nicely, and crater histories of the Moon and Mars indicate that early in its history (before 3.9 billion years ago), the inner solar system was pummeled again and again by gigantic meteors. Earth could not have escaped this drubbing. Indeed, Norman Sleep of Stanford University long ago proposed that life's only refuge on the primitive Earth would have been hydrothermal vents in the deep seafloor. Thus, the deepest branches of the tree may tell us of both evolution and extinction when life was young.)

As implied earlier, the Tree of Life provides a road map for history of life, its branching order reflecting the successive radiations of biological diversity. The tree suggests that early ecosystems were centered on hydrothermal vent and spring systems, with the later appearance of photosynthesis enabling life to spread across the planet. Large, complex organisms like plants and animals are evolutionary latecomers, confined to distal twigs on a eukaryotic branch formed mainly by microscopic organisms.

There is another way to interpret the tree. Because organisms in general, and microorganisms in particular, are commonly tied to specific habitats, the tree can be read as an environmental history of the Earth. For example, most early branching organisms do not use oxygen in metabolism, and many are killed by exposure to O_2 at even part-per-million levels. Organisms that can thrive when oxygen is present in moderate amounts branch later, and only at the tips of the tree do we find organisms like ourselves that require oxygen in high concentrations.

The Tree of Life, thus, makes predictions about Earth history that can be tested against the geological record. The first essential point of the tree is that the organisms and environments of our common experience are relatively recent features; the deep history of life is microbial. The other main point is that life has not evolved on a static planetary surface. Rather, life and environments have evolved together throughout our planet's history, inexorably linked by the biogeochemical cycles in which both participate.

Armed with predictions from comparative biology, we can turn our attention back to the Cambrian Explosion, captured so vividly in the cliffs along the Kotuikan River. The Tree of Life supports Charles Darwin's intuition that the Cambrian radiation of animals must have been preceded by a long antecedent history of life. Paleontologists wishing to reconstruct this history must focus on rocks deposited *before* the Cambrian period—on *Pre*cambrian rocks that document Earth's early planetary development. We also need to replace zoological search images by pictures drawn from microbiology. But bacteria, archaeans, and simple eukaryotic microorganisms are tiny and fragile. Can we really expect them to have left an interpretable fossil record?

3 | Life's Signature in Ancient Rocks

Sedimentary rocks on the arctic island of Spitsbergen formed 600–800 million years ago, well before the Cambrian Explosion. These rocks contain no traces of animal life, but viewed under a microscope, they teem with tiny fossils of cyanobacteria, algae, and protozoans. More conspicuous are stromatolites, reeflike structures built by microbial communities. Most pervasive, however, are chemical signals imparted by microbial metabolisms. Such discoveries encourage us to explore much older beds for evidence of life's earliest evolution.

SPITSBERGEN, an isolated and forbidding island halfway between Norway and the North Pole, is a starkly beautiful study in gray and white. The white is glacial ice that blankets most of the island; gray in various shades stripes the rocks that rise from the ice in great cliffs (figure 3.1). Tundra wildflowers add a splash of color, but vegetation is sparse. Century-old willows on coastal lowlands stand only a few inches above the ground, overtopped even by pale tufts of reindeer moss. Higher in the mountains only fairy rings of lichen, tiny but fiery orange, brighten the landscape. Seals and walruses sprawl languorously on grounded pack ice, and miniature reindeer graze on the miniature plants. Polar bears ply this coast, as well, fat from alfresco dinners of seal and curious about unusual sights like bright yellow tents pitched on a hill. Paleontologists sleep lightly in Spitsbergen.

In the mountainous northeastern quarter of the island, valley glaciers, majestic rivers of ice that flow with inexorable power but almost im-

Figure 3.1. Proterozoic rocks of the Akademikerbreen Group exposed in the glaciated highlands of northeastern Spitsbergen. Each band of light or dark gray rock is about 1,000 feet thick.

perceptible speed, have cut deep incisions into the landscape—so deep, in fact, that the Empire State Building, placed on the ice, would barely peek above a valley rim. What am I doing here? I asked myself as I crawled across a cliff top, unable to walk upright in the late afternoon gale. Like the How did we get here? query in chapter 1, this question has a spectrum of possible answers. On that blustery afternoon in the Arctic, the question might well have been rephrased, Why don't I work on tropical reefs? Transient disgruntlement aside, my question has a literal answer: my field partner, University of Iowa geologist Keene Swett, and I were logging a stratigraphic section through a thick succession of upper Proterozoic rocks exposed in the Spitsbergen cliffs. But once again, there is another answer, one that addresses why as well as what.

The top of the sedimentary pile in this region contains Early Cambrian fossils, much like those discovered along the Kotuikan River. Below extends a vast geological tome, more than 20,000 feet thick, written in a tropical seaway 800 to 600 million years ago. Consistent with their stratigraphic position beneath the oldest Cambrian strata, these Spitsbergen rocks contain no skeletons, no compressed carcasses, no tracks and trails—no evidence of animal life whatsoever. This doesn't necessarily mean that animals were absent when these beds were deposited, but any animals that did exist must have been tiny creatures that made little impact on accumulating sediments. Of course, the Tree of Life tells us that animals were preceded in time by other organisms, so we might search these rocks not for clams and brachiopods but instead for the fossils of algae, protozoans, or even bacteria. Northeastern Spitsbergen is a nearly ideal place to ask paleontological questions about this earlier biology. That's what we were doing there, seeking insights into life and environments before the Cambrian Explosion.

Spitsbergen isn't the first place where Proterozoic fossils were found, and I certainly wasn't the first person to find them. That honor goes Elso Barghoorn, the father of Precambrian paleontology and my mentor at Harvard. In 1954, Barghoorn and geologist Stanley Tyler reported the discovery of bacterial cells in nearly 2-billion-year-old rocks of the Gunflint Formation, western Ontario—I was three years old and unaware that the seeds of my professional life were being planted. Spitsbergen *is*, however, where my own odyssey in Precambrian research took wing. In 1978, a freshly minted Ph.D. and vividly green assistant professor, I was casting about for a project that would establish my scientific independence from the great Barghoorn. I had read about this inhospitable but potentially rewarding island in geological reports by Brian Harland of Cambridge University, and a hopeful letter to Harland ("Can you give me some pointers on working in Spitsbergen?") drew the best possible response: "Can you join us next summer?" Sure I could, little dreaming that this would redefine my research future.

The following summer found me on a small Norwegian fishing boat, sailing with three Cambridge shipmates around the northern tip of the island. Some mornings the sea was like glass, mirroring the bright sun as well the ice-covered peaks that lined the coast. Other days brought storms, and with them gray-green waves that tossed our boat (and my stomach) about like a toy. I knew little about sailing, less about wielding a pike to nudge ice out of our path, and absolutely nothing about what to do when whales approached our boat. (Slow down and steer *very* carefully.) I did know a little about paleontology, and, fortunately, promising rocks were abundant in the outcrops we studied.

A minor frustration of Proterozoic paleontology is that the quarry is usually too small to be seen by eye. We can only collect rocks that experience tells us might yield fossils, ship them optimistically back to the lab, and then, months later, examine prepared specimens to learn of success or failure. I was lucky. The first sample I looked at turned out to be brimming with exceptionally preserved microfossils. Peering down my microscope, I felt like Howard Carter beaming his lamp into King Tut's tomb, privileged to see "things, beautiful things" in the paper-thin slice of rock beneath my lens. I had a project, and over the next seven years I would return repeatedly to this island, traveling by helicopter and snowshoe in a protracted effort to learn the paleobiological secrets of this remarkable place.

I began by writing that the Spitsbergen rocks were deposited in a tropical seaway. How do we know this, and if we are correct, what are they doing on a mountaintop near the North Pole? Sedimentary geology is a vast repository of anecdote, lent structure by theory and experiment. A field geologist may notice the pattern of ripples that forms in sands along a lakeshore. Another will ask questions about these ripples in the laboratory, using a flow tank to determine the range of physical conditions under which they can form. Through repeated iterations of observation and experiment, an association grows between sediment pattern and the processes that govern its formation. In consequence, an experienced geologist can examine the pattern of bedding, structure, and texture in an ancient sandstone and from it infer the set of environments and sedimentary processes in play when it formed.

Interpreting the Spitsbergen rocks requires that we log the composition, thickness, and bedding features of successive layers. We must put hand lens to rock and eye to hand lens, scrunching our faces against the cliff to gain a magnified view of the ancient sediments. Every now and again, we wield a "persuader," the steel-shafted hammer tucked into every geologist's belt, to gather fist-size samples that can be shipped home. Back in the lab, field observations are supplemented by studies of thin sections (those paper-thin slices of rock) that reveal micron-scale features under the microscope. In the end, we associate each bed with a suite of characters that can be compared against features seen by other geologists in other rocks and in sediments accumulating today. Dialing into that collective pool of experience, we reconstruct the history of a small part of the world as it existed long ago.

The backbone of the Spitsbergen succession is the Akademikerbreen Group, a thick (seven thousand feet) stack of limestones and related rocks deposited near the edge of an ancient ocean (figure 3.1). Irregularly laminated dolomites[1] mark the shoreward edge of deposition. The millimeter-thick layers, or laminae, in these rocks (figure 3.2) closely resemble structures formed today where microbial mats spread across tidal flats, trapping and binding fine sedimentary particles to form thin layers, like delicate leaves of puff pastry. Prism-shaped cracks form a network of polygons in some beds. Most of us have seen similar features forming today where exposure to the sun causes wet muds to dry and crack; ancient mud cracks formed the same way.

Another curious feature helps to focus environmental interpretation. Here and there, sets of laminae several centimeters thick turn upward to form ridges, commonly fractured at their crests (figure 3.2). In cross section, these features resemble wigwams, hence, their easily remembered name: tepee structures. Tepees form today along warm lime-rich coastlines in the zone just above high tide, where surface sediments are baked by the sun and only occasionally inundated by the sea. In this regime, the growth of carbonate and gypsum crystals builds up pressure in surface beds. Eventually, the beds buckle, forming the cracked ridges.

The stacked beds of the Akademikerbreen Group, thus, provide a record of time *and* environment. We just learned that some of the Akademikerbreen carbonates formed at the very edge of a coastal carbonate platform. Not surprisingly, these rocks occur in close association with beds deposited just seaward, in the zone between high and low tides. Intertidal rocks are characterized by alternating sandy and muddy layers that record variations in wave and current energy, microbial mat layers without tepees, thicker beds formed by flowing tides whose angled, sandy laminae impart a herringbone texture to the rock, and shallow channels cut by tides into underlying sediments. A comparable

[1] Limestone consists of calcium carbonate ($CaCO_3$) particles cemented together to form a rock; the closely related mineral dolomite [$CaMg(CO_3)_2$] also forms rocks, called dolomite or (especially in the British Isles) dolostone. Most dolomites seen in the geological record formed by the chemical alteration of sediments that were originally limy.

Figure 3.2. Akademikerbreen carbonates deposited near the high-tide mark show wavy laminations characteristic of cyanobacterial mats, as well as tepee structures that provide a key to environmental interpretation. Nodules of black chert within the carbonate beds contain abundant fossils of filamentous microorganisms. Scale is 6 inches long.

suite of features can be seen today in places like the Bahama Banks, where they form a clear sign of tidal-flat sedimentation.

Intertidal rocks, in turn, mingle with beds formed below the low-tide mark, in a coastal lagoon filled by lime muds, sands, and flakes stripped from the tidal flat during storms. The ancient Spitsbergen lagoon was protected by a shoal of ooids, tiny spheres of concentrically laminated carbonate whose modern counterparts form where coastal waves repeatedly suspend particles in warm, lime-rich waters. Laminated domes and candelabrum-like structures up to several feet thick also punctuate the succession—patch reefs built by microbial communities.

The first lesson learned by aspiring geologists is "the present is the key to the past." Thus, modern sediments on the Bahama Banks help us to make sense of the Akademikerbreen Group. The ability to link processes observable today with patterns that can be recognized in ancient rocks makes the geological elucidation of our planet's history possible. We must be careful, however, not to get carried away by the beauty of this premise. Uniformity of process does not boil down to *plus ça change, plus c'est la même chose*. The tectonic, sedimentary, and geochemical *processes* at work today may have been in force throughout Earth history, but that doesn't mean that the *state* of our planet's surface has been invariant through time. Ocean chemistry, geography, and climate have all changed through time in ways that have been decisive for the history of environments—and life.

Mies van der Rohe, the great Bauhaus architect, reputedly said that "God is in the details." That's also where we'll find the keys to ancient states of the ocean and atmosphere. Take, for example, the ooid shoals mentioned earlier. Today, marine ooids have a maximum diameter of about one millimeter—the size of sand grains. The Akademikerbreen ooids, in contrast, reach the size of garden peas. Evidently, the chemistry of the Spitsbergen seaway wasn't quite like that of its closest modern analogues; it was more highly charged with calcium and carbonate ions, causing ooids to accrete faster and attain larger sizes than can be accomplished today. The giant ooids of Spitsbergen provide a first hint that the Precambrian Earth was not simply our own world with the plants and animals stripped away, an observation that will be developed in later chapters into a principal theme of deep Earth history. For now it is sufficient to remember that the *uniformitarian principle*—"the present is the key to the past"—is a statement about process, and one that should be viewed more as working hypothesis than universal truth in studies of the early Earth.

Having outlined our reasons for interpreting the Spitsbergen rocks as products of tropical sedimentation, we should consider, at least briefly, why they sit today in refrigerated cliffs north of the Arctic Circle. The explanation is that plate tectonic processes transported them to their current position. The hypothesis that continents have drifted through time was proposed early in the twentieth century by the German meteorologist Alfred Wegener, but it gained wide acceptance only in the 1960s and 1970s, when geophysical observations revealed how a seafloor conveyor belt, formed at oceanic ridges and destroyed beneath deep trenches, moves continents from place to place. Northeastern Spitsbergen moved poleward through the Paleozoic and Mesozoic eras, reaching its current latitude more than 100 million years ago. Then, with the opening of the Atlantic Ocean, this piece of real estate broke away from its closest geological relatives (now in Greenland), eventually to enter a deep freeze as the great Pleistocene Ice Age began. The geographic repositioning of Spitsbergen is a bit of good fortune, leaving us with rocks that are beautifully exposed and little altered by surface weathering—an unlikely prospect had this landmass remained at low latitudes.

By now, we know that the Spitsbergen rocks formed before the dawn of the Cambrian, in coastal environments at the edge of a tropical ocean. But was there life in that ocean, and did it leave a record in the Akademikerbreen sediments? That's what we really want to know, and to find out, we must search for rocks likely to preserve delicate biological remains. Chert (also known as flint) is one such rock, an extraordinarily hard substance made up of tiny interlocking crystals of quartz (crystalline silica, or SiO_2). Chert is hard enough to withstand the mechanical ravages of tectonic deformation and impermeable enough to shield its contents from corroding fluids. Encased in chert, then, sedimentary features—including *biological* features—can be preserved for the ages.

Chert is common in tidal-flat deposits of Precambrian age, often occurring as black nodules within carbonate beds (figure 3.2). The nodules formed within the sediments, not on the seafloor, as demonstrated by lamination and other features of bedding that run unbroken through the silica and surrounding carbonate. What's more, the cherts display textural features normally associated with lime deposits—they contain the same ooids, microbial mats, and crystalline cement textures found in adjacent carbonate rocks. In many cases the nodules formed soon after deposition, before burial initiated the compaction that bent encompassing sediments around them. Silica has no color of its own; the blackness of the nodules comes from included organic matter.

Spitsbergen cherts contain abundant and remarkably well-preserved microfossils—exquisite tiny gems locked in a tomb of silica. Cherts in carbonates that formed above the high-tide mark usually contain only a single type of microfossil, thick-walled tubes about 10 microns across that form a tightly woven fabric in the rock (plate 2a). (A micron is extremely short—one-thousandth of a millimeter, or forty-millionths of an inch. An eyelash is more than ten times as wide as these fossils.) The tubes are interpreted as the extracellular sheaths of filamentous cyanobacteria (plate 2b), those hardy bacterial practitioners of "green plant" photosynthesis. The microbial weave indicates that these minute organisms formed the microbial mats whose signature is written in the wavy lamination of encompassing carbonates. Low-diversity cyanobacterial mats occur today along the shoreward edge of restricted embayments from the Florida Keys and the Bahamas to the Persian Gulf and the arid coast of Western Australia.

On present-day tidal flats, microbial diversity increases toward the ocean, and Spitsbergen rocks show the same pattern. A series of mat-building cyanobacteria-like populations subdivided the ancient tidal gradient, forming discrete communities of mat builders and dwellers (organisms that lived in but did not contribute to the formation of the mat—like clams that nestle among frame-building corals in modern reefs).

Proterozoic microfossils have long been compared with living cyanobacteria, but how close is the comparison? Most "blue-greens" have simple shapes, and the morphological similarity between ancient and modern forms might mask deep physiological differences. Do we really mean to imply that cyanobacteria found today evolved before trilobites graced the oceans? One beautiful Spitsbergen population provides unusual insight into this issue. *Polybessurus bipartitus* consists of spheroidal cells 10 to 30 microns in diameter atop stalks made of extracellular secretions (plate 2c). Along the seaward edge of the ancient tidal flat, *Polybessurus* fossils are found as isolated individuals, but in more frequently exposed areas, they occur in dense populations that formed patchy crusts on the sediment surface. As my then graduate student Julian Green (now at the University of South Carolina) first recognized, morphological variations in the preserved fossils allow us to reconstruct their life history. Cells settled on the tidal-flat surface and, as they grew, began to secrete a series of extracellular envelopes. The stalks formed by successive envelopes enabled cells to maintain their position at the sediment-water interface despite an influx of lime mud. Once individuals reached a certain size, they divided repeatedly without intervening growth to form small cells that dispersed and settled again on the sediment surface, beginning the cycle again.

That's a lot to know about a Precambrian microfossil, enough for us to seek meaningful comparisons with living organisms. Frustratingly, published compendiums of cyanobacterial biology do not describe living populations with the suite of characters observed in the Spitsbergen fossils. But we knew something else about our microfossils; they lived along a tidal flat that bordered a subtropical to tropical seaway where carbonate sediments accumulated.

Armed with this knowledge, I traveled with my friend, academic neighbor (at Boston University), and cyanobacteria guru Steve Golubic to the closest modern environmental analogue we could identify—the Bahama Banks. (Science occasionally compensates those who summer in Spitsbergen.) There, on the lonely western edge of Andros Island, we found small black crusts dotting a tidal flat of lime mud laced with cyanobacterial mats. The crusts formed in the upper part of the intertidal zone, built by small spheroidal cyanobacteria that secreted extracellular sheaths elongated in the downward direction (plate 2d). That's right. Here, in a place predicted by Proterozoic rocks, we found the modern counterpart we sought—living but hitherto undescribed cyanobacteria whose morphology, life cycle, and environmental distribution match the ancient *Polybessurus bipartitus*.

The Spitsbergen example isn't an isolated instance of good luck. Steve's former graduate student Assad Al-Thukair (now at King Faisal University in Saudi Arabia) has discovered a half dozen new species of cyanobacteria that bore into and live within ooid grains; nearly all have exact fossil counterparts in silicified Proterozoic ooids of Spitsbergen and eastern Greenland. Because these cyanobacteria display stereotyped patterns of boring, even behavior can be included in the list of features shared by living and fossil populations. Steve and Montreal University's Hans Hofmann have drawn equally fine-scale ancient-modern comparisons between mat-building cyanobacteria found today on arid tidal flats and fossils that lived in comparable environments 2 billion years ago.

Collectively, these discoveries put some teeth into the old saw that many Proterozoic fossils look like cyanobacteria. Because habitat range is a direct function of physiology, the close environmental similarity between ancient and living cyanobacteria suggests that the microorganisms distributed across Spitsbergen (and other Proterozoic) tidal flats were essentially modern in morphology, life cycle, *and* physiology. Many of the cyanobacteria we see today are indeed survivors from the ancient Earth.

Cyanobacteria are common today in coastal habitats where very salty water or other environmental challenges restrict invasion by animals. By coincidence, the chert nodules in Proterozoic carbonates also center on coastal environments where silica was precipitated much like salt from evaporating seawater. Thus, chert's paleontological lantern shines most brightly on just those environments where cyanobacteria have always thrived. Cyanobacteria do not live alone on modern tidal flats, however; mat communities contain a host of other organisms, especially bacteria. Why don't we see this greater microbial diversity in chert nodules?

Tidal flats are harsh environments. At low tide, their inhabitants must tolerate the desiccating glare of a blazing sun; salty water provides an osmotic trial when the weather is dry; fresh water does the same during storms. Cyanobacteria respond to these challenges by secreting an extracellular envelope that protects the cells inside. That envelope is of particular importance to paleontologists because, unlike the cells within, it resists bacterial decay after death. Cyanobacteria, then, have the microbial equivalent of a clamshell, and in tidal-flat cyanobacteria this feature is especially well developed. While other bacteria live on tidal flats, most lack preservable walls or envelopes. And to make matters worse, they are tiny and have simple shapes that frustrate biological interpretation. The very fact that preserved fossils show evidence of post-mortem decay means that heterotrophic bacteria must have lived in tidal-flat environments. As discussed below, geochemical signatures enable us to identify at least a few of these populations, but we must face the fact that what we see preserved in thin sections of chert, while extraordinary, represents only a limited sampling of the microorganisms that lived along the Proterozoic shoreline.

Fortunately, the sample preserved best is one well worth understanding. Cyanobacteria are the working-class heroes of the Precambrian Earth—the main primary producers in early oceans and the source of the oxygen that transformed terrestrial environments. We know a great deal about living cyanobacteria, including their phylogenetic relationships. Add the good fortune that they are easily preserved and include species that can be recognized by form alone, and it becomes clear that cyanobacteria make an excellent flagship for paleontological studies of early life.

Although most fossils in the Spitsbergen cherts are undoubted or probable cyanobacteria, rare specimens of relatively large microfossils (greater than 100 microns), some shaped like miniature vases (plate 2f) and others studded with spines, provide tantalizing glimpses of a different biology. These fossils are limited to sediments that washed into tidal channels from offshore, suggesting that if we continue our environmental transect seaward, we might discover a diversity of Proterozoic life barely hinted at by the cherts.

During our field expeditions I collected many samples of black shales that accumulated in quiet subtidal environments beyond the ooid shoal. (Like the cherts, these shales are black because they contain organic matter.) Mesmerized by the chert biotas, I didn't do much with our shale samples, but when Nick Butterfield (now, like Harland, at Cambridge University) came on board as a graduate student, I suggested that he take a look at them to get some firsthand experience with Precambrian rocks.

Microscopic fossils are common in shales of all ages, packed cheek by jowl with clay minerals that inhibit decay. The mineral fabric of these rocks can be dissolved in strong acid, leaving organic remains to be mounted on glass slides and studied by optical or electron microscope. Conventional preparations of Spitsbergen shales yield conventional fossils, but Nick developed a set of unconventional procedures that enabled him to identify and gently free fragile remains. His painstaking work revealed a paleontological treasure trove. Plenty of cyanobacteria occur in these shales; neither then nor now are these microbes limited to tidal flats. But the Spitsbergen shales also contain diverse fossils of eukaryotic organisms, recognizable by their distinctive shapes. The vase-shaped fossils that washed onto the tidal flat are there, as are the large cells decked with spines. More exciting, however, the shales contain multicellular algae, the remains of small seaweeds that formed lawns on the shallow seafloor. Some of these fossils closely resemble green algae that can still be seen today (plate 2e). Others, however, have no close modern counterparts. Like trilobites and dinosaurs they are extinct, consigned by selection or catastrophe to the dustbin of (natural) history.

The Spitsbergen fossils are abundant and beautifully preserved, they are distributed over a range of sedimentary environments, and they include both prokaryotes and eukaryotes. On the other hand, they occur only in limited horizons of black chert and shale. The true ubiquity and diversity of late Proterozoic life is revealed by other biological indicators, most conspicuously *stromatolites*, the wavy-laminated, domed, and candelabrum-like structures seen in Akademikerbreen rocks (figure 3.3).

Stromatolites are the predominant features of carbonate rocks formed in Precambrian oceans. Stromatolitic buildups are uncommon today, but examples from places like the Bahama Banks and, especially, remote Shark Bay in Western Australia show how they form. Communities of microorganisms spread across sediment surfaces, weaving together to form coherent mats. Cyanobacteria (and, sometimes, algae) at the mat surface trap and bind fine particles supplied by waves and currents. As a veneer of mud or sand accumulates, these populations grow upward, reestablishing the mat at the sediment surface. Deeper in the mat, bacteria consume dead cells, changing local chemistry in a way that causes carbonate crystals to form. The processes of colonization, trapping and binding, and carbonate precipitation are discontinuous, but endlessly repetitive, with the result that fine layers of limestone accrete, one atop another. Stromatolites can be planar, domal, conical, or cylindrical; each one records a history of microbial growth on the ancient seafloor.

Chert nodules in the wavy laminated carbonates of Spitsbergen tidal flats preserve a direct record of mat-building microorganisms, and in one locality, the cyanobacterial architects of an offshore stromatolitic reef were preserved by fine carbonate cements that encrusted individual filaments. Most Spitsbergen stromatolites, however, contain no microfossils and so must be interpreted as biological features by invoking the

---

Figure 3.3. Stromatolites in the Akademikerbreen Group. (a) A microbial patch reef, some 15 feet thick, seen in a cliff face. (b) Close-up of a columnar stromatolite, showing the characteristic pattern of convex upward lamination. Note pocketknife for scale.

(a)

(b)

association between sedimentary pattern and microbiological process outlined in the preceding paragraph. That's not such a bad practice in younger Proterozoic successions like Spitsbergen, but as we shall see, assumptions about stromatolite formation become more contentious as we descend deeper into the past.

In general, then, stromatolites provide a sedimentary proxy record of microbial communities, revealing, like Friday's footprints in the sand, the presence but not the character of their makers. Still, this information is helpful, for it shows that 600–800 million years ago, microorganisms colonized nearly every available surface on the Spitsbergen seafloor, from tidal flats to the open ocean.

Like stromatolites, organic matter is much more widely distributed than microfossils in Akademikerbreen rocks. Roger Summons, an Australian geochemist now at the Massachusetts Institute of Technology, has shown that the organic matter in Proterozoic sedimentary beds includes *biomarkers*, biological molecules of known origin that are preserved in and can be extracted from the rock. These molecules consist mainly of lipids that have survived a gauntlet of bacterial decay. (Sadly, nitrogen- and phosphorus-rich molecules like DNA have a vanishingly small likelihood of preservation in very old rocks.) To date, the search for bio-markers in Spitsbergen rocks has not been notably successful, but else-where, especially in shales of similar age exposed deep within the Grand Canyon, abundant and diverse biomarker molecules preserve molecular signatures of archaeans, bacteria, protozoans, and algae, most of which have left no recognizable microfossils.

Biology is encrypted in the Spitsbergen rocks in one additional, even more generalized way. Individual microorganisms are tiny, but their col-lective physiological effect can be strong enough to influence the chem-ical composition of the ocean. As a prime example, photosynthetic or-ganisms influence the *isotopic* composition of carbonate minerals and organic matter deposited beneath the sea.

Isotopes provide our second set of Jacob Marley facts (the metabolic diversity of bacteria was the first). We have to come to grips with this bit of chemistry, because isotopes will allow us to track aspects of metabolic evolution through time. Moreover, as we'll see in subsequent chapters, isotopes provide a key to understanding the interplay between life and environmental change through our planet's history.

Carbon atoms come in three varieties, distinguished by molecular weight. About 99 percent of all carbon is in the form of ^{12}C, meaning that it contains six protons and six neutrons, for a total molecular weight of twelve. (Electrons have negligible mass.) Most of the remaining 1 percent consists of ^{13}C, its extra neutron contributing to a molecular weight of thirteen. There is also a bit of ^{14}C (two extra neutrons), but this form is radioactive and decays to nitrogen on a timescale of millennia. Thus, ^{14}C doesn't figure in discussions of very old rocks.

Because of their differing molecular weights, these *isotopes* behave differently in some chemical reactions. Notably, when photosynthetic organisms take up carbon dioxide to form organic molecules, CO_2 containing the lighter variety, ^{12}C, is incorporated more readily than CO_2 that contains ^{13}C. In consequence, the ratio of ^{13}C to ^{12}C in organic matter made by photosynthesis will be distinctly different from that of carbonate minerals formed in the same environment, a quantitative difference known as *fractionation* (figure 3.4). The difference is not large—about 25 to 30 parts per thousand—but it can easily be detected by geochemists armed with mass spectrometers. And this fractionation is preserved in sediments, providing us with a geochemical probe for ancient photosynthesis. (Organisms like ourselves that eat plants, algae, cyanobacteria, or other photosynthetic bacteria do not impart much additional fractionation in the process.) In the Spitsbergen rocks, the ratios of carbon isotopes in carbonates and organic matter consistently differ by about 28 parts per thousand—so photosynthesis fueled ecosystems in late Proterozoic oceans, just as it does today.

Chemistry also provides a paleobiological probe for sulfate-reducing bacteria. As noted in chapter 2, sulfate reducers play a key role in completing the marine carbon cycle, using sulfate ions (SO_4^{2-}) to respire organic molecules. The sulfate is converted to hydrogen sulfide (H_2S) that may combine with iron to enter the sedimentary record as pyrite (FeS_2)—the fool's gold sold in rock shops. Biological sulfate reduction shows a chemical preference for ^{32}S (16 protons and 16 neutrons) over the heavier isotope ^{34}S (two extra neutrons), resulting in sedimentary pyrite that is enriched in ^{32}S relative to gypsum formed from the same water body. Spitsbergen rocks indicate that the essential biological

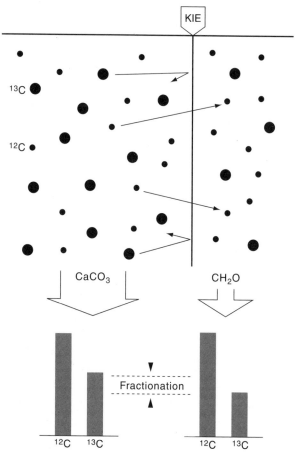

Figure 3.4. Diagram illustrating how photosynthetic organisms fractionate carbon isotopes. Black dots on left side of the diagram depict carbon dioxide molecules that contain ^{12}C (smaller) or ^{13}C (larger). Photosynthetic organisms fix $^{12}CO_2$ preferentially, with the result that the organic matter in photosynthetic organisms (and the organisms that eat them) is depleted in ^{13}C relative to its surroundings; biochemists speak of this as a kinetic isotope effect—hence, the label "KIE" in the figure. The isotopic *fractionation* imparted by organisms will be preserved in sediments as the difference in the ratio of ^{12}C to ^{13}C between limestone and organic matter in the same sample.

components of the sulfur cycle, like those of the carbon cycle, were in place when the gray rocks of this arctic island accumulated.

As observed on first sighting, the frozen Proterozoic rocks of Spitsbergen contain no bones, shells, or fossil trackways—nothing that would reward the casual collector (or Darwin!) on a weekend fossil hunt. But the apparent lack of fossils is deceptive—shells simply provide the wrong search image for Precambrian paleontology.

All of the carbonate minerals and organic carbon in the thick Spitsbergen succession bear the isotopic imprint of photosynthesis, and sulfur-containing minerals similarly preserve a metabolic signature of sulfate-reducing bacteria. Stromatolites document the ubiquity of microbial communities on the seafloor, while microfossils record aspects of biological diversity on the seafloor as well as in the water column.

When we look carefully, then, the fingerprints of biology are all over the Proterozoic rocks of Spitsbergen. The geological record *does* contain a record of early evolution that can be used to trim the Tree of Life. Our experience in Spitsbergen tells us how to approach ancient rocks and what to look for. But at 800 million years, the oldest beds on this desolate island are still relatively young. What happens when we apply these lessons to the bottom of the pile?

4 | The Earliest Glimmers of Life

At 3.5 billion years old, sedimentary rocks of the Warrawoona Group, Western Australia, provide one of our earliest glimpses of life and environments on the young Earth. Warrawoona rocks contain stromatolites and microscopic structures interpreted as fossil bacteria, but these interpretations remain contentious. Chemical signatures provide more convincing evidence of life's antiquity, although the type of biology they record is also uncertain. In geological investigations of Earth's earliest life, we still look through a glass darkly.

JAMES HUTTON, the late-eighteenth-century father of geology, wrote in challenging prose, but he did manage one epigram familiar to every Earth scientist. The geological record, Hutton observed, shows "no vestige of a beginning, no prospect of an end." Prospects of an end still seem remote, but over the past two decades, paleontologists have uncovered what can genuinely be regarded as vestiges of life's beginnings.

On a sunny day in late July, I am off to North Pole to examine these vestiges. Heat and dust permeate the cab as our Land Rover rattles over the rutted dirt track. There are flies everywhere. This North Pole, you see, lies in northwestern Australia—its name, with characteristic Aussie humor, marking one of the hottest places on Earth (figure 4.1) Bumping along in the passenger seat, I listen idly to Frank Sinatra on the radio as I try to read the geological landscape around us. It isn't easy. Trained to discriminate among shades of gray in arctic outcrops, my eyes fail me in the Australian bush, where everything is tinted red. For-

Figure 4.1. These low hills near North Pole, Australia, are built of sedimentary and volcanic rocks formed nearly 3.5 billion years ago. North Pole rocks preserve some of our earliest evidence of life and environments on the young Earth. Note Land Rover for scale.

tunately, I'm in good hands. At the wheel is Roger Buick, at the time a Harvard postdoctoral fellow and now professor of geology at the University of Washington. A brilliant iconoclast whose wiry frame and unruly mane camouflage the sophisticated scholar within, Roger has eyes of unparalleled geological keenness in these low rubbly hills.

Covered only sparsely by needle-sharp spinifex grass and scraggly acacias, the North Pole hills expose an extraordinary remnant of the early Earth, a thick succession of volcanic and sedimentary rocks called the Warrawoona Group that formed nearly 3.5 billion years ago. Folded and compressed between ovoid domes of granite, these rocks have for the most part been profoundly altered by metamorphic heat and pressure. Only at North Pole and a few other spots has some tectonic grace preserved them in a little-modified state. This is the place to ask questions about the antiquity of life.

Before asking such questions, however, we need to address an issue that was glossed over in our discussion of Spitsbergen rocks—proof of age.

How do we know that the North Pole rocks formed more than 3 billion years ago?

Geologic time can be measured in two ways. Any event discernible in the rock record can be used to divide Earth history into three intervals: the time preceding the event, the time of the event itself, and all subsequent time. By mapping the distribution and spatial relationships of rocks seen locally, a series of events can be placed in order of occurrence to form a *relative* timescale. In principle (though not always in practice), the rules are straightforward. Sediments, ash spewed from volcanoes, and lava flows settle by gravity onto the land surface or seafloor. For this reason, any such layered rock will be younger than the beds it covers. Volcanic rocks that intrude into other units must be younger than the rocks they pierce. And events that alter rock accumulations—folding, faulting, erosion, and metamorphism—obviously must postdate deposition.

Knowing these simple relationships, we can work out the relative timing of events evident in a road cut, quarry, or mountainside. For example, in the geological cross section cartooned in figure 4.2, the oldest identifiable event is the deposition of bed *A*, followed sequentially by the formation of beds *B* through *F*. Later, these beds were folded and, after that, intruded by granite *G*. Then, after erosion planed off these older units, beds *H* through *J* were laid down. The fault on the right-hand side of the figure cuts all older sedimentary units and, therefore, must have occurred late. The youngest event we can infer from the cross section is the erosion that sculpted the current land surface.

Extension of such a local history to the global scale requires a means of establishing the time relationships of rock units found in different regions, say, in the Rockies and the Appalachians, or in North America and Australia. For rocks deposited since the Cambrian Explosion, fossils provide our best guide to stratigraphic correlation. Indeed, the eras, periods, and finer subdivisions of the geologic timescale (page 2) reflect more than anything else the changing composition of life through time. Sedimentary and volcanic rocks may also preserve distinctive features of chemistry or magnetism that complement or, in some cases, substitute for correlations based on fossils.

Fossils can reveal that two rocks units are the same age, but they cannot tell us what that age is. For that, we need a natural chronometer ca-

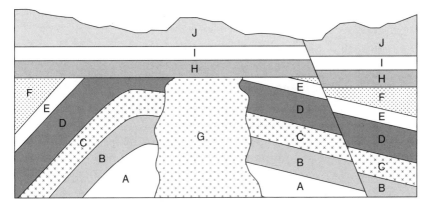

Figure 4.2. A geological cross section, illustrating how geologists sort out relative age relationships. See text for discussion.

pable of recording quantitatively the passage of time. Radioactive isotopes bound into rock-forming minerals provide geology's clocks.

Radioisotopes are inherently unstable atoms that decay spontaneously to stable daughter elements at rates that can be measured accurately in the laboratory. This being the case, if we can determine how much of a radioactive parent element has disappeared from a mineral through time, or how much stable daughter has accumulated, we can calculate the age of the mineral itself. Interestingly, what remains constant in radioactive decay is the *proportion* of radioisotopes that decay in a set interval, not the number of atoms. Thus, as the abundance of a radioactive isotope in a mineral decreases through time, the absolute rate at which the isotope decays also declines. The pace at which a radioactive isotope decays is called its *half-life*—the time it takes for half of the radioisotope in a material to decay to another element.

Classically educated readers will be reminded of Xeno's paradox, the ancient Greek puzzler in which Achilles chases a hare. Being a hero, Achilles runs faster than his prey, halving the distance between hunter and hunted each minute. When will Achilles catch the hare? The answer of course is never, because if the hare is moving at a constant speed, Achilles must continually be slowing down. Starting at a 200-yard disadvantage, he may race 200 yards in the first minute to the hare's 100, but in minute two he'll cover only 150 yards, and by minute four, he'll make a mere 112.5 yards. Note that if we know how far Achilles has run and how his speed varies with distance from the hare, we can figure out how long he's been pursuing his frustrating chase. This, in essence, is how radiometric dating works.

The best-known radiometric dating system is provided by ^{14}C, or carbon 14, a rare isotope of carbon produced naturally by cosmic rays and anthropogenically by nuclear bombs. It decays to nitrogen (^{14}N) with a half-life of 5,730 years. Because ^{14}C is so uncommon (it makes up fewer than one in every thousand carbon atoms) and its half-life so short, radiocarbon dating is limited to the past 100,000 years or so. In older materials, there simply isn't enough ^{14}C left to measure accurately. In consequence, ^{14}C provides a great tool for Egyptologists or paleontologists interested in woolly mammoths, but it cannot help unravel Earth's deep history.

To date the Warrawoona succession, we need a more stately clock—a radioisotope whose half-life is measured in many millions or even billions of years. Potassium 40 (^{40}K) was recognized early on as a promising candidate for geochronology. This unstable isotope breaks down to form either calcium (^{40}Ca), which unfortunately can't be differentiated from calcium ions already in the mineral, or argon (^{40}Ar), which can. The half-life of ^{40}K is 1.25 *billion* years. Further, potassium is abundant and widely distributed in rock-forming minerals—it occurs in the feldspars that tint granites pink, in the microscopic minerals of volcanic ash, and in clays that form during weathering.

Despite these advantages, the potassium-argon chronometer is not much used by geologists interested in the early Earth. If ^{40}K behaves like a clock, tectonic and metamorphic processes act like toddlers eager to play with the dial. Geologic events that take place long after mineral formation can drive argon from minerals, resetting the clock and destroying the chemical memory of elapsed time. (An inert gas, argon is held only loosely within the chemical lattices of minerals.)

What we really need to date old rocks is a system that acts like the "black boxes" in airplanes—an isotope not easily lost from a mineral not readily altered. Zircons—uranium-bearing minerals found in granites and related volcanic rocks—are the flight recorders of Precambrian geology. In fact, the uranium bound into zircon crystals at their time of formation provides *two* reliable chronometers: ^{238}U decays to lead 206 (^{206}Pb) with a half-life of about 4.5 billion years (the age of the Earth), while the rarer isotope ^{235}U breaks down to ^{207}Pb with a half-life a bit longer than 700 million years. This provides a valuable cross-check of measured ages—if the two clocks don't give the same age, the zircon has been altered.

If zircons have a problem, it is that they are *too* tough. Unlike most other minerals, zircons can go through the entire rock cycle, from crystallization in an igneous rock to metamorphism and subsequent erosion to form a sediment grain, without loss of chemical integrity. Indeed, magma ascending through the Earth's crust can pluck zircons from surrounding rocks, incorporating older minerals (and, therefore, clocks) into younger rocks. What's more, zircons can grow during each passage through the Earth's interior; Archean[1] zircons may display a half dozen layers around a central core, each the accreted product of a specific geological event.

William Compston of the Australian National University developed an ingenious instrument for plumbing the radiometric complexities of ancient zircons. Called the Sensitive High Resolution Ion Microprobe (SHRIMP, for short—years ago, a fire in Compston's lab prompted the predictable flurry of "throw a little SHRIMP on the barbie" jokes), this instrument uses fine ion beams to sample the individual growth layers of zircons so that geochemists can date each one independently. The SHRIMP revolutionized Archean geology, allowing geologists to unravel the complex time relationships of sedimentary, volcanic, and tectonic events on the early Earth.

Now we can understand why the basal Cambrian rocks along the Kotuikan River are said to be about 543 million years old. Locally, sedimentary rocks bearing the earliest Cambrian fossils are interbedded with a volcanic unit that contains zircons; U-Pb chronology shows that the zircons crystallized 543 ± 1 million years ago. (The "± 1" is an estimate of the error associated with the measured age; it is a statistical statement which indicates that the true age of crystallization has a 95 percent chance of falling within the interval 542–544 million years. Good

---

[1] *Archean* (the time interval before the Proterozoic) and *Archaea* (a major branch on Tree of Life) both derive from the Greek *archaios*, meaning "ancient." Beyond that, however, the terms have nothing to do with one another. The similar names do not imply that the Archean Eon was the Age of Archaea. It's just that geologists and biologists don't talk much to each other.

geologists pay careful attention to those error bars.) There are no well-dated rocks in the Spitsbergen succession, but fossils and chemical features of the Akademikerbreen Group permit at least broad correlation with better-dated rocks found elsewhere.

All of which brings us back to the age of the Warrawoona Group. SHRIMP analyses of zircons in volcanic rocks near the top and base of the group yield ages of 3,458 ± 2 and 3,471 ± 5 million years, respectively. Knowing these ages so precisely is a triumph of Earth science that imparts special importance to studies of Warrawoona life and environments.

In the late Precambrian rocks of Spitsbergen, biological signatures can be found almost everywhere—in microfossils, in stromatolites that catch the eye wherever we look, in biomarker molecules preserved in sedimentary organic matter, and in the isotopic abundances of carbon and sulfur in rocks throughout the succession. What does the paleobiological exploration of Warrawoona rocks reveal?

The sedimentary/volcanic succession at North Pole is, in fact, mostly volcanic and only a little sedimentary, not a good omen for the paleontologist. More than 95 percent of the unit consists of lavas poured out on land or in shallow water, along with ash layers and coarser beds made of fragmented volcanic rocks. Three and a half billion years ago, Warrawoona might have looked something like the volcanic necklace of the Indonesian Archipelago. In detail, however, Archean geography resists characterization by simple analogy to modern features.

Sediments, mostly preserved as dark cherts, accumulated in coastal basins nestled among the volcanoes. By now we know that chert lifts the paleontological heart, but the silica-rich rocks of North Pole formed in a manner quite different from those in Spitsbergen. These older cherts precipitated from volcanically heated fluids that percolated through Warrawoona sediments, replacing original minerals soon after deposition. Unfortunately for the paleontological heart (and mind), chert formed this way is as likely to destroy biological signatures as preserve them. Further complicating the issue, at least some of the cherts occur as veins that fill cracks in the sedimentary and volcanic pile. They appear to be part of a hydrothermal plumbing system, much like the network of underground conduits that feed present-day hot springs in Yellowstone Park.

Unraveling the depositional history of these rocks requires that one carefully map the distributions of constituent rock types and then use outcrop and laboratory observations to peer through the silica veil. Roger Buick did just that as a doctoral student at the University of Western Australia. Along with his Australian colleagues, Roger demonstrated that most Warrawoona sediments originated as muds, sands, and cobbles eroded from surrounding volcanoes and deposited in a flanking basin. From time to time, sandbars or other barriers blocked the basin's mouth, restricting the flow of water into coastal lagoons. Evaporation increased the concentration of dissolved calcium and carbonate ions in these waters, leading to the production of calcium carbonate as whitings—millions of tiny crystals that formed in the water column, turning the lagoon milky white before they settled onto the seafloor as lime mud. Rosettes of gypsum crystals decorate many beds, deposited as evaporation proceeded still further. Actually, the gypsum in these rocks is long gone; its former presence is recorded by silica ghosts that preserve gypsum's distinctive crystal form (figure 4.3a).

Perhaps the most unusual rocks in the North Pole region are beds of barium sulfate, or barite—fans and more continuous layers of slender prisms that grew like so much rock candy on the seafloor. Barite is uncommon in younger sedimentary successions, but Wouter Nijman and his colleagues at Utrecht University in the Netherlands believe that the large mounds of barite in the Warrawoona Group formed where hot hydrothermal fluids erupted onto the seafloor. From a human perspective, these eruptions seem alien and inhospitable, but to the heat-loving microorganisms found on early branches of the Tree of Life, they would have furnished a hot microbial Eden.

In the field, we can see wavy laminations in sedimentary rocks, much like the microbially laminated carbonates observed in Spitsbergen. In a few places, these laminations are bowed upward to form domes or cones—the geometric hallmarks of stromatolites (figure 4.3b). Warrawoona stromatolites were first reported in 1980 by Stanford University's Don Lowe and, independently, by Australia's Malcolm Walter, John Dunlop, and, of course, Roger Buick. The analogy with present-day stromatolite growth, noted in the preceding chapter, was pushed to the limit: in some of the oldest sedimentary rocks on Earth, these geologists discerned a familiar signature of biology.

(a)

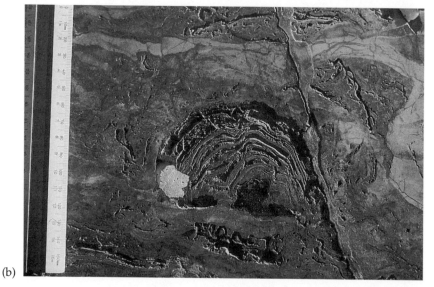

(b)

Figure 4.3. Sedimentary features of Warrawoona rocks. (a) Gray-black gypsum crystals (now replaced by silica) that grew on the seafloor and were subsequently buried by thin beds of mud and sand (light layers). Scale bar in centimeters. (b) The stromatolite discovered by Roger Buick and colleagues in the early 1980s, now subject of much debate. Six-inch scale to left. (Photo (a) courtesy of Roger Buick)

Or did they? Early on, Roger Buick and his colleagues urged caution: the words "possible" and "probable" pervade a thoughtful essay published in 1983. It isn't, they wrote, that the Warrawoona structures *can't* have been built by bacteria, but rather that the actual evidence for biological accretion is iffy. And, in rocks this old, with so little independent evidence of biology, the interpretational stakes are high. We want to know whether or not life existed 3.5 billion years ago, and "maybe" is not a satisfying conclusion.

In 1990, Don Lowe retreated further. He explicitly reinterpreted Warrawoona stromatolites in terms of chemical processes that deposited layers of minerals on the seafloor and physical processes that deformed these layers into upturned folds, like carpets after a slip. In this view, the hand of life is nowhere to be seen.

Why this loss of confidence in a biological interpretation? Simply put, microbial mat processes can give rise to laminated structures, but they are not the *only* processes that can do so. Similar features can form without microbial mats if ambient waters are highly charged with dissolved minerals—for example, in Yellowstone Park, where silica-saturated waters spill episodically from thermal springs to form laminated structures that resemble stromatolites built by microorganisms.

We didn't worry much about this in Spitsbergen, because 600–800 million years ago, the oceans weren't sufficiently laden with calcium and carbonate (or silica) to drive this type of deposition. On the early Earth, however, ocean chemistry was different. John Grotzinger, an MIT geologist we'll meet again, has shown that Archean oceans were charged with calcium and carbonate ions, so much so that the carbonate minerals calcite and aragonite commonly formed by direct precipitation on the seafloor. These deposits include giant crystal fans whose precipitated origins are obvious, but they also encompass flat laminated beds, domes, and small columns arrayed in parallel rows (see chapter 7). This being the case, it becomes important to examine the microscopic features of Warrawoona stromatolites to see whether mat-building fossils have been preserved. They have not, making it difficult to know to what extent organisms participated in the formation of these structures.

At present, the interpretation of Warrawoona stromatolites remains unresolved. We can accept them as biological only if we can rule out alternative physical explanations. Hans Hofmann and Kath Grey, two experienced stromatolite hands, recently discovered new Warrawoona structures that they hope will decide the case in favor of biology. The buildups, formed layer after layer by the precipitation of thin crystalline veneers, are conical, a shape rarely generated on the seafloor in the absence of microorganisms.

But if the Warrawoona cones reflect life's guiding hand, what *kind* of biology was at work? On the modern Earth, cyanobacteria are the best-known mat builders, and so there is a tendency to associate all ancient stromatolites with cyanobacterial mats. Other bacteria can form mats, however, and there is no a priori reason to assume that cyanobacteria existed 3.5 billion years ago. In our quest to find biological signatures in the ancient beds of Warrawonna, stromatolites provide only a suggestive scrawl—a tantalizing but ambiguous hint that life began in our planet's infancy.

Spitsbergen cherts contain microfossils that are unmistakably cyanobacterial, increasing our confidence that these blue-green bacteria built contemporaneous stromatolites. Are the black cherts of Warrawoona similarly informative? For more than a decade, most people thought so, but recent reinvestigations have sown considerable doubt about Warrawoona micropaleontology.

In a preliminary draft of this chapter, I recounted the 1987 discovery of Warrawoona fossils in cherts from a knobby outcrop along the dry bed of Chinaman's Creek, near North Pole. In these rocks, UCLA's Bill Schopf and Bonnie Packer found tiny filaments 1 to 20 microns in diameter and up to a few hundred microns long (figure 4.4). The structures are rare and poorly preserved—the distorting effects of crystal growth are clear in published photos. Nonetheless, the photos, or at least the interpretative drawings that accompany them, look like simple cyanobacterial filaments. Equally, however, they resemble other types of bacteria, limiting taxonomic or metabolic interpretation.

In Spitsbergen, knowing where a microbe grew helped us to understand how it made its living. Geology colors our thoughts about Warrawoona biology, as well, but in a surprising way. Careful mapping by Australian stratigrapher Martin van Kranendonk shows that the cherty rocks at Chinaman's Creek formed beneath the Warrawoona seafloor, not on it—these cherts originated in hydrothermal

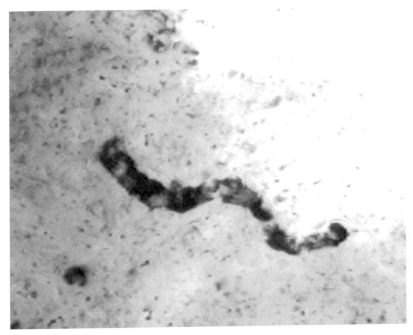

Figure 4.4. Microstructures interpreted as bacterial fossils in Warrawoona chert. The alternative interpretation is that they are simply chains of crystals formed in a hydrothermal vein. (Photo courtesy of Martin Brasier)

veins, as outlined earlier in this chapter (figure 4.5). Conceivably, cyanobacteria could have found their way into these cracks along with other sedimentary particles, but the environmental setting revealed by geology likely favored chemosynthetic growth over photosynthesis.

From here, the problems of interpretation mount. Despite reservations about the specific interpretation of Warrawoona microstructures, I had no cause to doubt their biological origins as I drafted my chapter. After all, poorly preserved fossils with shapes obscured by mineral growth occur throughout the geological record. Why should the old cherts of Warrawoona be any different? Before committing my opinions to print, however, I wanted to see the Warrawoona material for myself.

Surprisingly, the place to view purported Warrawoona fossils is not Sydney or Perth, or even Los Angeles. It is London, in the collections of the Natural History Museum. In September 2000, I had to cross the

Figure 4.5. Warrawoona chert at Marble Bar, Western Australia. The red (pigmented by iron oxides, gray in picture) and white bands accumulated on the seafloor. In contrast, the black bands cut across other beds and so are younger—and formed in a different way. Warrawoona microstructures interpreted as fossils come from these crosscutting cherts, interpreted as hydrothermal plumbing systems filled by silica.

ocean for a scientific meeting in Oxford and so made arrangements to spend a quiet day at the museum before returning home. I also planned a not-so-quiet day in Oxford visiting Martin Brasier, a well-known paleontologist and connoisseur of local pubs. Fortunately, I told Martin of my hope to study the Warrawoona rocks, and he, in turn, told me that the key samples were out on loan—to Oxford! In Martin's lab in a quaint Edwardian building off Parks Road, we spent a revealing day examining thin sections of Warrawoona chert.

In Spitsbergen cherts, microfossils are abundant. They have shapes similar to those of living microorganisms, but different from shapes made by purely physical and chemical processes. Most retain at least some of their original organic matter. And some even occur in environmental settings much like those of close living counterparts. No comparable claims can be made about the Warrawoona microstructures.

Observed through the microscope in Martin's lab, the tiny filaments in Warrawoona rocks looked like minerals.

As a boy, I often spent idle summer afternoons gazing at clouds. Most were billowy masses, beautiful but shapeless. Every now and again, however, an unmistakable face appeared in the sky. Or a castle. Or a lion. For a moment, the shapes took striking form, but even as a youngster I was pretty sure that they were, in the end, just clouds. Are the Warrawoona microstructures "just clouds," as well?

That question is difficult to answer based on a few pictures cut and cropped for publication. It requires *context*—the framework provided by the overall rock fabric seen in thin sections of Warrawoona chert. It was the rest of the clouds that revealed my air castles as watery illusions, and it is the overall fabric of Warrawoona cherts that casts doubt on those rare features that look biological. Martin and his colleagues painstakingly documented how volcanic and hydrothermal processes shaped the cherts from Chinaman's Creek. They believe that physical processes can account for *all* microscopic features of the cherts, including those singled out as fossils. If this interpretation is correct, the Warrawoona microstructures cannot be cellular filaments, only stacked crystals that mimic but do not preserve a record of biology.

Are paleontology's crown jewels, so old and rare, made of paste? In fairness, Bill Schopf disputes this reading. In a rebuttal to the claim by Brasier and colleagues, Bill and University of Alabama chemist Tom Wdowiak show that the disputed Warrawoona structures contain organic matter at their margins. This, of course, is consistent with the view that they are microfossils, but it doesn't end the debate. Archean cherts commonly contain the ghosts of early formed minerals whose distinctive shapes are preserved by a veneer of organic matter. My own guess is that most Warrawoona structures are mineral chains draped by an organic film (that may, itself, have a biological origin). Continuing study may yet confirm the presence of fossils in these rocks—the debate is far from over—but I doubt that any such remains will teach us much about early ecosystems. Warrawoona microstructures, like Warrawoona stromatolites, can only suggest that something interesting and important lies just beyond our grasp.

Biomarker molecules are not retained in rocks from North Pole, but isotopic signatures are; the carbon and sulfur isotopes of Warrawoona rocks provide our best indication of life's deep history. As in Spitsbergen (and nearly everywhere else that sedimentary rocks were deposited in Precambrian times), $^{13}C/^{12}C$ ratios in Warrawoona carbonates and organic matter differ by about 30 parts per thousand. This difference is most easily explained by photosynthesis, but given our experience with stromatolites and microfossils, we should ask once more whether physical processes can mimic the effects of biology. Some chemical reactions do form organic molecules depleted in ^{13}C. Only under carefully controlled experimental conditions, however, does nonbiological fractionation approach the levels recorded in Warrawoona rocks. For this reason, the *consistently* large fractionation measured in North Pole samples suggests the presence of an early biosphere.

Carbon isotopes in organic matter from the chert veins that cut through Warrawoona sediments and lavas could record chemosynthetic bacteria that lived in hydrothermal waters. But the widespread distribution of organic matter in sedimentary rocks deposited *on* the seafloor supports the hypothesis that photosynthesis fueled microbial life in the Warrawoona ocean. Whether primary producers were mostly cyanobacteria or other types of photosynthetic bacteria with a similar isotopic signature remains uncertain. Sulfur isotopes in sedimentary pyrite and barite likewise suggest that sulfate-reducing bacteria lived the Warrawoona lagoon, although this, too, has been questioned by geologists grown skeptical about early Archean biosignatures.

At present, that's about all we can say. The heat-scorched hills of North Pole suggest that life existed 3.5 billion years ago, and that, by itself, is remarkable. Warrawoona communities may have included photosynthetic microorganisms and other microbes with metabolisms still seen today. But many uncertainties persist. Warrawoona paleobiology remains a shadow play whose apparently familiar themes may be deceptive.

Northwestern Australia is one of two places in the world that contain well-preserved sedimentary rocks as old as 3.5 billion years. The other is the rugged Barberton Mountain Land, near Kruger Park in South Africa. The two areas are similar, so much so that some geologists believe that they form parts of a single ancient terrain, severed by plate tectonic movements long after the Archean. Their paleontological inventories bear comparison, as well. Both include stromatolites of uncertain origin, along with organic matter whose carbon isotopic composition and sedimentary distribution suggest some type of photosynthesis. Both have been heated to temperatures that destroy biomarker molecules. And like those of Warrawoona, Barberton cherts contain spherical and filamentous microstructures reminiscent of fossils.

As a graduate student, I had a go at Archean paleontology, traveling to Africa as Elso Barghoorn's field assistant. Having grown up on Tarzan books, I was excited as the plane touched down in Johannesburg late at night. I couldn't wait to catch my first glimpse of Africa the next morning, and was only a little disappointed that the view from my hotel window looked a lot like Chicago. Within hours, we were on the road, and as the cityscape receded in the rearview mirror, the great South African veldt opened before us. Culturally, ecologically, and geologically, the Barberton Mountain Land was new to me. In each clump of thorn trees I sensed menace, and in each chert I espied fame. Neither fame nor menace materialized, but the cherts did turn out to contain microstructures that are probably, if not unambiguously, biological.

In one particular sample marked by centimeter-scale stromatolite-like precipitates, I discovered a large population of spherical microstructures 2 to 4 microns in diameter—the size and shape of small cyanobacteria (figure 4.6a). The structures occur in individual laminae. Moreover, they are made of organic matter, and some preserve both an outer wall and a raisinlike interior body, also organic. The microstructures are compressed along the bedding surface, much like younger microfossils—indeed, this slight flattening tells us that the structures formed before enclosing sediments were compacted by burial. The distribution of sizes in the population matches that of modern cyanobacteria, and the structures also show evidence of binary division, again much like living blue-greens.

So are these fossil cyanobacteria? Not necessarily. Many different bacteria are small and spherical. More sobering, nonbiological processes can, in principle, produce similar structures—although it isn't obvious that such processes were at work in the Barberton seaway. The Barberton spheres, thus, fall only a few steps ahead of Warrawoona filaments. They could be fossil cyanobacteria or some other type of microorganism. They

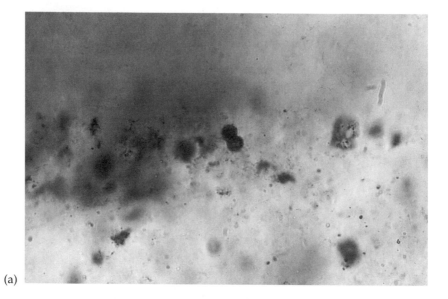

(a)

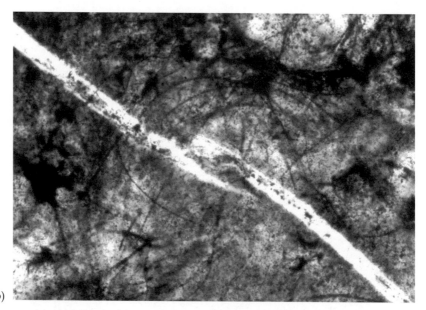

(b)

Figure 4.6. (a) Carbonaceous microstructure, possibly preserving a microbe during cell division, in 3.4-billion-year-old rocks from South Africa. Sphere is 4 microns in diameter. (b) Filamentous microfossils in 3.2-billion-year-old rocks from northwestern Australia. Each filament is about 2 microns across. (Photo (b) courtesy of Birger Rasmussen)

could record a primordial microbe that is long extinct. Or they could be carbonaceous spheres formed by physical processes on the Barberton seafloor. We simply don't know. More recently, Maud Walsh of Louisiana State University has made a careful study of the organic matter in Barberton cherts, finding bedding textures most easily explained as mats and thin filaments that may be microfossils.

What kind of planet can we piece together from these fragments? Geologically, it appears to have been a world of familiar processes but not-so-familiar patterns. Continents began to form at least 4.2 billion years ago, and chemical details of volcanic rocks from Barberton, Warrawoona, and other old terrains suggest that a large volume of continental crust had formed by the time they were deposited. Little of these early continents remains, however, implying that on the early Earth, continents were recycled back into the mantle more easily than they are today. Three and a half billion years ago, plate tectonics had already begun to pattern our planetary surface, but Earth's upper mantle appears to have been hotter, the basaltic crust beneath the oceans thicker, and, perhaps, the continents smaller and less stable. Then, as now, continental crust probably formed at plate margins, where descending slabs of oceanic crust cause overlying rocks to melt. On the other hand, early continent formation may have received a significant boost from a source that is no longer important—partial melting of basalts buried beneath thick piles of lava spilled onto the seafloor.

The rock record that survives from the early Earth is not simply the fragment of a geologically modern planet buffeted by time. Something about the character and mix of processes that form and destroy continents was different, and though many insightful scientists have hazarded opinions, we don't fully understand what it was.

We have a bit more confidence that when the Warrawoona seaway formed, Earth was a biological planet. Moreover, the evidence of carbon isotopes suggests that the great ecological liberation of photosynthesis may already have begun. Whether or not contemporary microorganisms included the oxygen-producing cyanobacteria is uncertain, but the presence of *any* type of photosynthetic organism in the Warrawoona ocean speaks volumes, because it allows us to place a calibration point on the Tree of Life introduced in chapter 2. In the new view of microbial evolution symbolized by the tree, photosynthetic organisms are relative latecomers that diversified long after the origin of life and the divergence of biology's principal domains. If Warrawoona organic matter was made by photosynthesis, then a great deal of evolution must already have taken place.

Microorganisms appear to have cycled carbon, sulfur, and nitrogen through early Archean ecosystems, just as they do today. There is no record of eukaryotes or archaeans in these oldest rocks, but then there aren't many fossils, period, and it would be hazardous to interpret the absence of evidence as evidence for absence. The branching pattern of the Tree of Life tells us that if photosynthetic bacteria lived in the Warrawoona sea, then at least some Archaea were almost certainly present.

We have one more constraint on early Archean biology. Consistent with our environmental reading of the tree, geological observations indicate that 3.5 billion years ago Earth's atmosphere contained nitrogen, carbon dioxide, and water vapor but little free oxygen. Most inferences about ancient environments are gleaned from subtle geochemical clues, but the sedimentary signature of oxygen limitation is flamboyant—bright red bands of cherty rock rich in the iron oxide mineral hematite (Fe_2O_3). Aptly named banded iron formation (BIF, for short), these rocks do not form in the present-day ocean. In fact, with one important exception, they haven't accumulated for the past 1.85 billion years. But for the first half of Earth history, BIFs were a standard component of marine sediments. The reason that BIFs do not form today is that iron entering the oceans immediately encounters oxygen and precipitates as iron oxide; as a result, iron concentrations in modern seawater are extremely low. The BIF in Archean sedimentary successions may have formed by the reaction of iron with oxygen. Or the iron may have been oxidized by photosynthetic bacteria or UV radiation that penetrated to the sea surface in the absence of an effective ozone shield. However they were precipitated, the BIFs tell us that on the early Earth, iron didn't get stripped away as it entered the sea. Instead, iron was readily transported in solution throughout the deep ocean. This could only happen if deep waters were free of oxygen, forcing us to conclude that the Archean atmosphere and the sea surface in contact with that atmosphere had much less oxygen than today. Just how much oxygen was present is contentious, but it couldn't have been more than about 1 percent of present-day levels and may have been much less. Under these conditions, aerobic respiration and chemosynthetic metabolisms that depend on molecular oxygen may have been limited or absent, depending on just how much oxygen was available.

Together, then, geology and microbiology suggest that early Archean oceans differed from those of younger intervals, containing much less oxygen but more iron. The early ocean may have been warmer than today's, but geology provides us with few real constraints on Archean climate. All we can say is that if photosynthesis was present, surface waters could have been no warmer than 74°C—the maximum temperature tolerated by photosynthetic organisms. The Tree of Life suggests that the earliest Bacteria and Archaea lived at higher temperatures, but this need not be interpreted to mean that the entire ocean was hot. Early heat-loving microbes might well have lived in hydrothermal environments like those inhabited by their descendants today.

How can we amplify the weak biological signal in early Archean rocks? Two strategies come readily to mind; one concerns how we look for evidence and the other, where we search. As the saga of Warrawoona illustrates, the paleontological search strategy that has worked so well in younger rocks—collect black chert, and lots of it—has not been notably successful in early Archean rocks. Malcolm Walter, a leading Australian paleontologist, has advocated a different kind of search, one that focuses on the very hydrothermal processes that ravaged the contents of preserved cherts. As noted earlier, hydrothermal systems are home to some of the earliest branching organisms on the Tree of Life; moreover, hydrothermal springs commonly deposit carbonate and silica minerals that might preserve records of early biology. Chimneys of pyrite and other minerals also form at some hydrothermal vents in the ocean. Comparable deposits occur in early Archean terrains, but until recently they have not received much attention from paleontologists. This situation is changing, however, and early in 2000, Australian geologist Birger Rasmussen reported convincingly biological (if metabolically uninformative) filaments from hydrothermal mineral deposits that, at 3.2 billion years, are not much younger than Warrawoona (figure 4.6b). I anticipate that as our search strategy widens, our understanding of early biology will grow proportionately.

The second obvious research path is to discover older rocks. Because of our planet's restless surface, metamorphism, uplift, and erosion continually alter and destroy the rock record—successively older periods of time are represented by ever smaller volumes of rock. This being the case, it is no simple task to find little-altered sedimentary rocks older than the Warrawoona and Barberton successions. Yet, Roger Buick has done just that. In a remote area of northwestern Australia, he discovered a succession of sedimentary and volcanic rocks that lies *beneath* the Warrawoona Group. Christened the Coonterunah succession, these beds contain volcanic rocks dated at 3,515 ± 3 million years—not dramatically older than the Warrawoona rocks that cover them, but nonetheless a step deeper into Earth history. Coonterunah rocks include sedimentary beds deposited in deepwater environments along with basaltic lavas, but to date they have not yielded fossils.

Steve Mojzsis and his colleagues took a bigger step backward—to cirea 3.8-billion-year-old rocks on Akilia Island off the coast of southwestern Greenland. These rocks have been severely altered by metamorphism, making it difficult to read their geologic history. Mojzsis and colleagues interpret the rocks as sediments laid down on an ancient seafloor. Within the rocks, they found tiny grains of mineral phosphate, and, within these mineral grains, still tinier inclusions of reduced carbon (graphite). Using an ion microprobe, Mojzsis measured the isotopic composition of the carbon and found that it is strongly depleted in ^{13}C, consistent with formation by biological processes.

But nothing is ever simple in Archean geology. After careful reexamination of the Akilia rocks, a team led by phosphate expert Gus Arrhenius (one of the "and colleagues" in Mojzsis's original group) concluded that the phosphate grains formed relatively late in the history of these rocks, during alteration by hot metamorphic fluids. Moreover, Arrhenius and colleagues believe that the graphite in these grains formed at the same time, by chemical reaction of metamorphic fluids with iron carbonate in the rock. Independent research by geologists Christopher Fedo and Martin Whitehouse supports a metamorphic origin for key features of the Akilia rocks. Indeed, Fedo and Whitehouse believe that the rocks originated as igneous bodies deep within the Earth's interior.

Here's one more debate still to be resolved. But if the revised interpretation is correct, the carbon in Akilia rocks can tell us nothing about life. In fact, Arrhenius's conclusions undermine more than that. Recall that those graphite crystals in Akilia phosphate grains have carbon-isotope ratios much like organic matter produced by photosynthesis. If physical processes can fractionate carbon isotopes by up to 50 parts per thousand, then our faith in carbon-isotope composition as a biosignature is shaken.

Faith shaken, but also, perhaps, faith restored. Elsewhere in southwestern Greenland, Minik Rosing, of the Geological Museum in Copenhagen, Denmark, has found a succession of metamorphosed shales some 160 feet thick. The rocks are more than 3.7 billion years old, are assuredly sedimentary, and contain abundant particles of graphite distributed much like the organic matter in younger shales. And once again, carbon isotopes suggest biological activity. In this case, however, Rosing argues convincingly that his graphites formed by the heating of *organic* matter, not the alteration of older minerals. Biology provides the simplest explanation for the chemistry in Rosing's rocks—but a second line of evidence would be reassuring.

At present, our knowledge of Archean life and environments is both frustrating and exhilarating—frustrating because we are certain of so little, but exhilarating because we know anything at all. It is stimulating, as well, because the companion of ignorance is opportunity.

Some of the biggest questions focus on what came before Warrawoona, or Barberton, or even Akilia. If the oldest sedimentary rocks we can identify provide hints of complex microorganisms, what kinds of cells lived still earlier? Indeed, how did biology arise in the first place?

5 | The Emergence of Life

Life was forged by the same physical and chemical processes that shaped our planet's crust and oceans. Life is different, however, because it can undergo Darwinian evolution. Natural selection has played a key role in the evolution of plants and animals; early in our planet's history, it also directed the chemical evolution that made life possible. In general terms, we understand how biological molecules might have evolved from simpler precursors present on the early Earth. But how proteins, nucleic acids, and membranes came to interact so intricately remains a mystery.

I<small>N A WELL-KNOWN</small> travelers' tale, an explorer from the West treks up an Eastern mountain in search of a venerated sage. Finding the wise man in his aerie, and eager to show him a thing or two, the explorer asks, "What lies beneath the mountains we see around us?" "The mountains, the valleys, and everything else on the Earth rides on the back of a giant turtle," replies the wise man. "What, then, lies beneath the turtle?" continues the traveler, sensing that his host has taken the bait. "Why, another turtle," says the sage. "And beneath that?" "Another turtle." "And then?" "Yet another turtle." And so it goes, until the wise man, exasperated at last by his thickheaded guest, cries, "Don't you see? It's turtles all the way down!"

Biology has a "turtles all the way down" problem of its own. In *The Origin of Species*, Darwin hypothesized that new species arise by the modification of old—that the raw material of life is life. Louis Pasteur, Darwin's great Parisian contemporary, went a step further. In his decisive refutation of spontaneous generation, the long-held view that life can arise de novo from nonliving materials, Pasteur declared with Latin economy that *"omne vivum ex viva."* Life springs *always* from life.

In science, answers provoke new questions, and so it isn't surprising that in resolving two of biology's greatest conundrums, Darwin and Pasteur laid bare its most profound mystery. Perhaps life has sprung only from life for the past four billion years, but sometime, somewhere, in the earliest days of our planet, our first ancestors had to arise from something else.[1]

Rational thought about the origin of life actually predates both Darwin and Pasteur. For example, in 1804, before his famous grandson was even born, Erasmus Darwin captured the essence of biological history in verse:

Organic life beneath the shoreless waves
Was born and nurs'd in ocean's pearly caves;
First forms minute, unseen by spheric glass,
Move on the mud, or pierce the watery mass;
These, as successive generations bloom,
New powers acquire and larger limbs assume;
Whence countless groups of vegetation spring,
And breathing realms of fin and feet and wing.

The younger Darwin knew these lines as he pondered natural selection, and he may have recalled them again in 1871 when he wrote to Benjamin Hooker about the ultimate origin of species:

It is often said that all the conditions for the first production of a living organism are now present, which could ever have been present. But if (and oh! What a big if!) we could conceive in some warm little pond, with all sorts of ammonia and phosphoric salts, light, heat, electricity, &c., present, that a proteine [*sic*] compound was chemically formed ready to undergo still more complex changes, at the present day such matter would be instantly devoured or absorbed, which would not have been the case before living creatures were formed.

In that letter, in a half dozen lines of conversational prose, Darwin outlined the central idea that has guided scientific thinking about the

---

[1] Claiming that life on Earth was seeded from Mars (see chapter 13), whether probable or not, doesn't solve the problem; it merely shifts its location.

origin of life ever since. The energy of nature drives simple molecules to combine and recombine, building chemical complexity until a system emerges that can replicate itself. The idea is powerful and intuitively appealing—life, seemingly so distinct from water and rock, arose from the same planetary processes that shaped Earth's physical features. The question is how to test it.

The oldest sedimentary rocks we know of already contain at least a scrappy signature of biology, so we can't recover a direct record of life's origins from geology. The alternative is to devise laboratory experiments that allow us to evaluate the plausibility of hypothesized steps along the road to life. We can't know with historical certainty whether any specific reaction played a role in the emergence of organisms, but we can seek to understand, in general terms, how chemical reactions on the early Earth could have made biology possible.

Sugars were synthesized from formaldehyde precursors as early as 1861, but only with an ingenious experiment conducted by Stanley Miller in 1953 did experimental research on the origin of life take flight. Working in the University of Chicago laboratory of Nobel laureate Harold Urey, Miller asked whether lightning could have synthesized the raw materials for life as it ripped through the primordial atmosphere. Others had thought about this question—Russian chemist Alexander Oparin and British biologist J.B.S. Haldane had both written penetrating essays on the origin of life in the 1920s—but Miller did more than cogitate. He filled a glass vessel with a mixture of methane, ammonia, hydrogen gas, and water vapor—judged by Urey to approximate Earth's earliest atmosphere—and then repeatedly ran a spark though the vessel. Within a few days, the flask changed color, tinted reddish brown by a film on its inner surface. When Miller analyzed the fluid from which this gunk had precipitated, he found a variety of organic compounds, including amino acids, the building blocks of proteins.

In one remarkable experiment, Miller jump-started research on life's origins. Powered by the energy of nature, simple gas mixtures could give rise to molecules of biological relevance and complexity. Amino acids and other biologically interesting compounds occur in carbonaceous meteorites and do so in proportions strikingly similar to those generated by Miller. Thus, what happened in Miller's flask was not some esoteric reaction likely to occur only in the laboratory, but rather a chemistry found widely in our solar system and beyond.

But, as ever, the answers provided by Miller's simulation prompted new questions. Will any combination of primordial reactants do, or do biologically interesting molecules form only when you get the recipe right? Miller himself answered this one; the recipe matters a lot. Miller-Urey synthesis yields diverse and abundant organic molecules only when the ratio of hydrogen to carbon atoms in the gas mixture is at least four to one. That means that the chemistry in Miller's beaker could have been important on the early Earth only if the primordial atmosphere was strongly reducing—devoid of oxygen and rich in hydrogen, methane, and/or ammonia. As discussed in chapter 4, most people agree that oxygen was scarce when Earth was young, but beginning with UCLA geochemist William Rubey in the 1950s, many workers have also come to believe that our planet's earliest air was only weakly reducing, a mixture dominated by CO_2 and nitrogen gas rather than methane and ammonia. If the early atmosphere was indeed so weakly reducing, we need to look elsewhere for the kilns where biology's bricks were made. Where might we look? More fundamentally, what should we be looking for?

Modern cells share a number of key features (figure 5.1). All have membranes that line the cell's exterior and regulate molecular traffic into and out of the cytoplasm. Cells also synthesize proteins that catalyze chemical reactions or provide structural support.[2] And cells maintain a chemical library of information in the form of DNA. Critically important, membranes, proteins, and DNA continually interact within the cell. An impressive arsenal of proteins enables cells to consult the DNA library, replicate it in toto, or transcribe portions of it into RNA messages that provide blueprints for the formation of more proteins. The translation of RNA messages takes place in ribosomes, chemical factories made of protein and RNA woven together. Organisms also grow and reproduce, requiring that they take up materials and energy from their surroundings—metabolism employs still more proteins, some of them embedded in the cell's membranes.

---

[2] Proteins that catalyze chemical reactions are called *enzymes.*

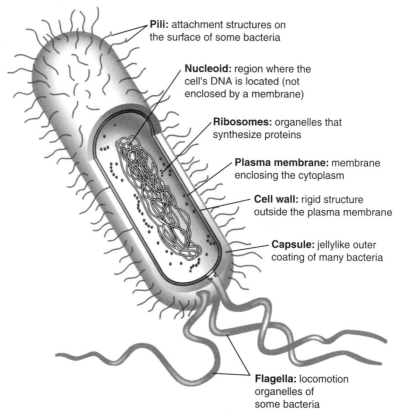

Pili: attachment structures on the surface of some bacteria

Nucleoid: region where the cell's DNA is located (not enclosed by a membrane)

Ribosomes: organelles that synthesize proteins

Plasma membrane: membrane enclosing the cytoplasm

Cell wall: rigid structure outside the plasma membrane

Capsule: jellylike outer coating of many bacteria

Flagella: locomotion organelles of some bacteria

Figure 5.1. The structure and function of a bacterial cell. RNA messages are transcribed from DNA within the nucleoid; these RNA messages are subsequently translated into proteins in chemical factories called ribosomes; cellular metabolism is carried out by pigments and proteins embedded in the cell's membranes. (Reproduced with permission from N. A. Campbell and J. B. Reece, *Biology*, Sixth Edition. Copyright © 2002 by Pearson Education Inc.)

Even the simplest living organisms, then, are tremendously sophisticated molecular machines. The earliest life-forms had to be much, much simpler. We need to think about a family of molecules simple enough to form by physical processes, yet sufficiently complicated to lay the evolutionary groundwork for living cells. Such molecules would have contained information and structure sufficient to replicate themselves and, eventually, to direct the synthesis of other compounds that could catalyze replication with increasing efficiency. And molecules able to initiate an evolutionary trajectory through which life could wean itself from the physical processes that gave it birth, synthesizing molecules needed for growth rather than incorporating them from the ambient environment, and tapping chemical or solar energy to fuel the workings of the cell.

RNA holds a special place in this line of thinking (figure 5.2). Long known for its functions in the translation of DNA into protein, RNA once seemed to be biology's handmaiden, a mere go-between in a molecular drama dominated by DNA and proteins. As early as 1968, however, Nobel laureate Francis Crick envisioned a much grander role for RNA in life's earliest history. "Possibly," he mused, "the first 'enzyme' was an RNA molecule with replicase properties." At the time it was written, Crick's speculation probably struck many as daft. By the mid-1980s, however, it proved prophetic.

Crick's prophecy was fulfilled in Thomas Cech's lab at the University of Colorado. In studies of the ciliate protozoan *Tetrahymena thermophila*, Cech and his students had found that RNA destined for use in ribosomes was modified between the time of its transcription and the moment it glommed onto ribosomal proteins. Somehow, molecular surgeons snipped away an unwanted segment of the RNA molecule and then neatly spliced the remaining pieces back together. What enzymes, Cech asked, catalyzed this reaction?

Cech's team began by purifying the unedited RNA sequence. Then they added it to a solution of proteins extracted from the ciliate's nucleus. Not surprisingly, cutting and pasting of the RNA proceeded just as it does in the cell—the molecular catalyst was somewhere in the beaker. All good experiments, however, include controls designed to make sure that observed results aren't produced by circumstances other than those being tested, so the team prepared additional test tubes with RNA but no proteins. And that's where the surprise came. When Cech inspected the controls, he found that the RNA editing took place *even when proteins were absent*. The control experiment, usually the most mundane part of laboratory research, led Cech's team to an electrifying conclusion: the RNA had excised the snippet by itself and spliced itself back together again. RNA could store information like DNA *and* catalyze reactions like proteins.

A

B

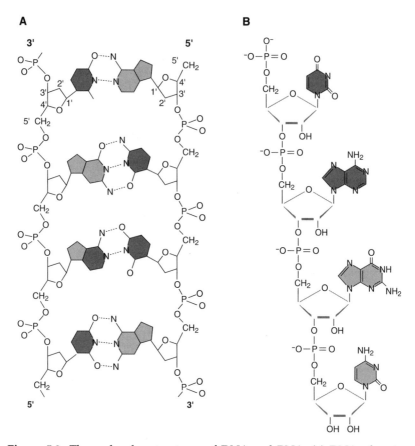

Figure 5.2. The molecular structures of DNA and RNA. (a) DNA, showing how a chemical backbone of phosphate and deoxyribose sugar combines with four bases that provide both molecular information and the bonds that link two strands into a double helix. (b) RNA, built from ribose sugar, phosphate, and four bases (one of which differs from its DNA counterpart). (a) Adapted from illustration by Irving Geis from R. E. Dickerson (1983). The DNA helix and how it is read, *Scientific American* 249: 97–112; rights owned by the Howard Hughes Medical Institute. Not to be reproduced without permission; (b) reproduced with permission from S. Freeman, 2002, *Biological Science*, Prentice Hall)

The discovery of RNA enzymes, or *ribozymes*, made independently at about the same time by Yale biochemist Sidney Altman, had—dare I say it—a catalytic effect on thinking about life's origins. As philosopher of biology Iris Fry put it, this remarkable molecule emerged as "both chicken and egg" in the riddle of life's origins. In 1986, Walter Gilbert, a Harvard colleague, crafted a short but stimulating essay whose title, "The RNA World," came to symbolize a conceptual way station in the evolution of biochemical complexity. As Gilbert saw it, RNA was the information-rich molecule that could form by self-organization and catalyze its own replication. Later, as life matured, evolution introduced a division of labor, with double helices of DNA providing a more stable library, while intricately folded proteins took over most catalytic functions.

This road to biology is tremendously appealing, but it contains a number of speed bumps. First and most important is the difficulty of making RNA under plausible prebiotic conditions. RNA molecules contain a backbone of the five-carbon sugar ribose and phosphate (PO_4^{3-}) joined together in a chain (figure 5.2). Four bases, chemical compounds built from rings of carbon and nitrogen, attach to the sugars, imparting molecular information. The bases can be synthesized easily enough—in 1961 Spanish biochemist Juan Oró showed that one of them, adenine, could form directly by combining five molecules of hydrogen cyanide (the modus operandi in many a detective story and likely present on the young Earth). Ribose, on the other hand, isn't so easy to explain. As noted earlier, sugars can be synthesized from solutions containing formaldehyde (probably also present in our planet's infancy), but ribose is only one of many products, and a minor one at that. The processes by which this sugar might have been thrust onto prebiotic center stage are not obvious. Worst of all, even if we could produce the right components, combining them to form nucleotides, the building blocks of nucleic acids, is daunting. To date, no one has figured out how to do it.

There is still another difficulty. Nucleotides are chiral molecules, which is to say that they come in two forms that are mirror images of each other—like your hands. RNA can be built from right-handed or left-handed nucleotides, but mixed chains won't grow. How, then, could RNA—which in cells consists exclusively of right-handed nucleotides—have emerged from a fifty-fifty mixture of left- and right-handed building blocks? Again, no one knows.

The problems are so difficult that many researchers have given up on the idea that RNA was the primordial molecule of life. They suggest instead that prebiotic evolution began with molecules without "handedness" that are easier to synthesize and polymerize. Nonchiral molecules that form double helices like those of nucleic acids can indeed be

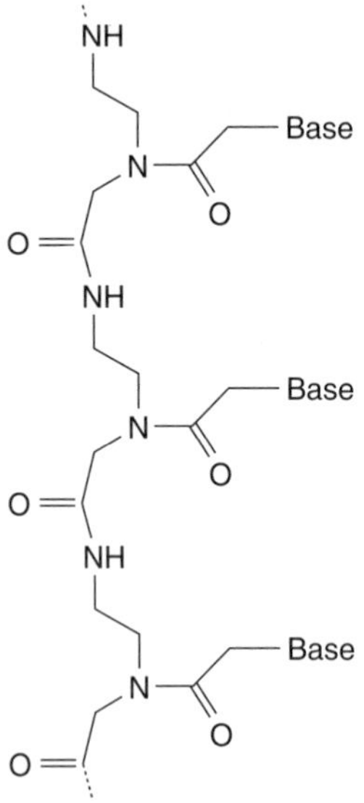

Figure 5.3. The molecular structure of peptide nucleic acid, a nonchiral molecule, illustrating one possible route in the evolution of nucleic acids.

generated with relative ease in the laboratory. Moreover, at least one of them, called peptide nucleic acid (figure 5.3), can direct the formation of its RNA complement, supporting the hypothesis that RNA could have replaced its primordial precursors later, in the course of evolution.

We have a better sense of how the RNA world might have operated once it burst onto the scene. Pioneering experiments in the laboratories of Jack Szostak and Jerry Joyce, at Harvard Medical School and the Scripps Research Institute, respectively, show how natural selection could have honed the function of RNA molecules. The experiments typically begin with millions of RNA strands generated randomly in the laboratory. Those showing weak ability to catalyze a specific reaction are selected and replicated repeatedly under conditions that introduce mutations into the replicates. A second round of selection ensues, followed by more replication. Repeated replication and selection yield functionally efficient ribozymes.

These experiments show that many types of RNA catalysis are possible; RNA provides just the sort of jack-of-all-trades molecule needed to get biology going. They also suggest that natural selection can generate molecular order from disorder, and amplify weak biochemical function. If Szostak, Joyce, and their colleagues are on the right track, evolution is not only the hallmark of biology, but a prerequisite for life.

This, in turn, highlights a cardinal feature of nucleic acid replication at the dawn of life. To quote Ronald Reagan, "mistakes were made." As early RNA molecules copied themselves, errors crept in so that molecular daughters included sequence variations not found in their parents. This variation supplied the raw material for chemical evolution on the early Earth, and it has fueled biological evolution ever since that time.

As in many other spheres of life, the Goldilocks rule applies. If the error rate of RNA (and later, DNA) replication is too high, successful variants can't be perpetuated in succeeding generations. If, on the other hand, it is too low, evolution cannot continue. That actual error rates are "just right" may seem a remarkable coincidence, but it isn't—it results from natural selection at the molecular level. Limited sloppiness is an evolutionary virtue.

We can, thus, envision a central role for RNA in nascent life. Experiments show that in the presence of mineral catalysts, nucleotides can join together to form RNA (although nucleotides, themselves, have not yet been built from scratch), and relatively short RNA molecules can direct their own replication. From the pool of sequence variation created by copying errors, natural selection can amplify those sequences that function best, those that replicate themselves a bit faster or with fewer errors than their molecular neighbors.

It turns out that much the same can be said for proteins. As Stanley Miller demonstrated, amino acids form readily under at least some prebiotic conditions, and, like nucleic acids, they can join together to form peptides, the amino acid chains that fold to form functioning proteins. Reza Gadhiri and his colleagues at the Scripps Research Institute have even generated peptides that catalyze their own replication. Thus, depending on environmental conditions, nucleic acid *and* protein precursors could have evolved in primeval oceans.

Regardless of which (if either) set of molecules came first, however, the most formidable problem in primordial evolution must be the emergence of systems in which proteins and nucleic acids *interact*, each ensuring the survival of the other. Freeman Dyson, a renowned physicist who has thought deeply about life's origins, posited that life actually began twice, once via the RNA route and again by way of proteins. Cells with interacting proteins and nucleic acids subsequently arose by protobiological merger. The idea isn't crazy. As will be clear by the end of this book, innovation by alliance is a major theme in evolution.

Viewed as a chicken-and-egg problem, Dyson's proposal has obvious attractions. But the issue is more complicated than that, because it must also be approached as a question of locks and keys. Molecular cross talk between nucleic acids and proteins is mediated by the genetic code, a set of chemical correspondences that allows the molecular language of nucleotides to be translated into the amino acid chains of proteins. Was the code a "frozen accident," as suggested by Francis Crick, and if so what was the nature of the accident? Alternatively, do chemical rules underlie the molecular correspondences, and if so, what are they? The origin of the genetic code, and with it the emergence of biochemically complex life, remains biology's mystery of mysteries.

One further riddle remains to be explored. Metabolism weaned life from the physical processes that gave it birth, and the metabolic pathways introduced in chapter 2 have perpetuated biology for some 4 billion years. How does the evolution of metabolism fit into the scenario outlined in previous paragraphs?

Membranes made of phospholipids (molecules with a "head" of phosphate and organic carbon and two long "tails" of fatty acids) serve both to separate cells from their physical surroundings and to direct the metabolic traffic of ions, molecules, and energy. Contemporary phospholipids may, like DNA, have arisen during the course of early biological evolution; however, simpler molecules able to assemble spontaneously into membranous vesicles may well have existed in the primeval ocean; spherical membranes made from lipidlike compounds in meteorites look more than a little like cells.

Possibly, the linkage of metabolism and replication began with the packaging of RNA (or protein and RNA) molecules inside primitive membranes. A simple experiment conducted by University of California biochemist David Deamer illustrates how this might have occurred. Deamer took a mixture of DNA and lipid vesicles and repeatedly wet and dried it. When the mixture dried, the molecules formed a layered "sandwich" on the bottom of the reaction flask. During subsequent wetting, the lipids reconstituted their spherical vesicles, but now, some of the DNA strands were *inside* the vesicles. This suggests that the structural association of lipids, proteins, and nucleic acids could have arisen spontaneously on the early Earth. Moreover, Deamer and his colleagues were able to synthesize RNA within vesicles, using nucleotides imported across the bounding membrane. This provides a faint glimmer of how metabolism and replication could have become linked. And once again, the Goldilocks rule is in order.

Membrane function depends strongly on the length of the fatty acid "tails" in constituent phospholipids. If they are too short, the membrane will be so leaky that it won't work. If they are too long, nothing can pass through the membrane, an equally fatal circumstance. Thus, the first membranes, at least the first ones that worked, must have formed spontaneously with lipid "tails" long enough to keep large polymers on the inside, but short enough to allow smaller molecules to pass in and out of the vesicle.

The key to metabolic integration must, of course, have been the evolution of nucleic acid sequences that coded for proteins able to direct membrane synthesis. Once the membrane came under cellular control, it, too, became subject to natural selection, leading (again) to molecular division of labor. The phospholipid "tails" grew longer, prohibiting passage of all but a few molecules, such as water and simple gases. At the same time, proteins became embedded in the phospholipid matrix, providing specialized gates and channels that admitted ions, molecules, and energy in a controlled fashion. Like nucleic acids, then, membranes appear to have evolved from simple, unspecialized structures formed by chemical processes to sophisticated, specialized systems constructed by the cell.

The scenario outlined in previous paragraphs begins with nucleic acids or their molecular forebears and expands to include proteins, membranes, and ultimately, metabolism. Some scientists, however, believe that things happened the other way around—that life *began* with metabolism and subsequently invented nucleic acids and proteins.

Gunther Wächtershäuser, a Munich chemist and patent lawyer, has argued the case for life's metabolic origin with particular clarity and vigor. Conventional prebiotic syntheses, he notes, work only when environmental conditions are just right. Considering that conditions may not have been right in the primordial oceans, Wächtershäuser concludes that life must have gotten started some other way in some other place. The place he advocates is a hydrothermal spring, like those found in the Warrawoona seaway or along present-day midocean ridges. In these settings, hydrogen sulfide emitted from vents can react with iron mono-sulfide to form pyrite—great chimneys of pyrite still form at the mouths of deep-sea hydrothermal vents. The reaction yields both energy and chemical reducing power (in the form of hydrogen), powering, in Wächtershäuser's scenario, the fixation of carbon dioxide (or carbon monoxide) to form organic compounds on the surfaces of growing pyrite crystals.

Could life have begun as a film on fool's gold? We don't know, but recent laboratory experiments lend support to at least parts of Wächtershäuser's hypothesis. Among other things, Wächtershäuser and his colleagues have generated acetic acid by the chemical fixation of carbon monoxide on a slurry of iron and nickel sulfides. These slurries also catalyze the formation of peptide chains from activated amino acids. Many aspects of the metabolism-first hypothesis remain to be tested, and the scenario still faces the formidable problem of integrating metabolism with nucleic acids and proteins. The experiments do, however, suggest that we should approach unconventional hypotheses about prebiotic evolution with an open mind. Mao Zedong may not have meant it when he urged fellow revolutionaries to "let a thousand flowers bloom," but in research on the origin of life, we need all the ideas we can muster.

Once genes, proteins, and membranes were in place, life probably climbed the trunk of Darwin's great Tree of Life quickly, propelled by natural selection, gene duplication, and the lateral transfer of genes. Biological expansion required that genes take control over many functions, but it isn't necessary to believe that all genetic takeover occurred in one cell line. More likely, biochemical innovations arose individually in a number of distinct lineages. One line may have made vitamin B, another fatty acids, and a third proteins that catalyze replication. In the leaky world of protocells, the gene products of one cell may have been available to all, leading to complex communities linked by biosynthetic codependence.

(We still live in such a world—you need your morning orange juice because your cells cannot synthesize vitamin C.) Leaky membranes would have permitted genes as well as gene products to pass from one cell to another, allowing diverse biochemical pathways to aggregate in a small number of lineages destined to spread across the world.

Early metabolisms must have been simple, employing generalized enzymes that catalyzed many reactions with low efficiency. With time, however, natural selection produced enzymes that were both efficient and specific. For example, the first bacteria to use sulfate in respiration undoubtedly did so poorly, thriving largely because nobody else could do it at all. At first, sulfate would have been harnessed using enzymes that also served other functions, but as new variants better able to reduce sulfate outcompeted their compatriots, selection for efficient sulfate reduction progressed rapidly. Of course, this selection would have come at a cost, because the better that enzymes functioned in their new role, the more poorly they catalyzed other reactions.

Thus, as selection honed enzymatic function, evolving cells would have needed a source of new genes. Lateral transfer helped, although gene sharing probably slowed as increasingly competent membranes evolved. A second source, of continuing importance, was gene duplication. Errors in replication can result in extra copies of genes, providing raw material for evolutionary innovation. Once duplicated, the two gene copies can respond to differing selection pressures, eventually giving rise to two enzymes with distinct functions.

It is relatively easy to see how selection could have driven the evolution of specific enzymes, but what about complex metabolic pathways that integrate the activities of many proteins? The molecular intricacy of photosynthesis rivals the anatomical complexity of vertebrate eyes, chosen by Darwin to illustrate how "organs of extreme perfection" could evolve. Darwin's creationist critics believed that the complexities of the eye must reflect intelligent design, but Darwin knew better. Among living organisms, he noted, we find a spectrum of photosensitive structures that runs from simple eyespots (pigment concentrations in single cells) to eyes with muscles, lenses, and optical nerves. All meet the functional demands of their bearers. Thus, Darwin argued that natural selection can fashion complexity from simplicity as long as all intermediates are functional.

The same is true of photosynthesis. At first glance, the photosynthetic apparatus may appear to be nature's most elegant molecular assembly, but closer inspection reveals it to be her most wonderfully contrived Rube Goldberg machine, its impressive complexity resolvable into a series of components, each with its own origin and evolution.

To begin with, chlorophyll, the central pigment in photosynthesis, appears to have evolved from simpler but functional precursors. The earliest of these were probably present in the prebiotic environment, whereas later transitional forms still occur as chemical intermediates in chlorophyll biosynthesis. As the biosynthetic pathway of chlorophyll evolved, each new step modified the function of the previous end product.

Chlorophyll, in turn, is linked with proteins in an intricate molecular assembly called a photosystem that turns sunlight into chemical energy (figure 5.4). Once again, present-day complexity appears to have evolved by gene duplication and lateral transfer. Gene duplication elaborated the families of proteins that transport electrons in photosynthesis, whereas the paired photosystems of cyanobacteria (and green plants) originated in separate groups of bacteria and came together by lateral transfer. (All the genes required to build and operate a photosystem occur together along a strand of DNA. Thus packaged, this functional cassette of genes appears to have migrated from one photosynthetic bacterium to another, perhaps aided by viruses or by uptake from dead cells.)

Figure 5.4. Diagram showing the molecular assembly of photosystems in cyanobacteria and green plants. Chlorophyll and other pigments absorb photons of light and transfer their energy to "excited" electrons. The electrons are then passed in bucket-brigade fashion along a chain of proteins embedded in the photosynthetic membrane. This set of chemical reactions culminates in the formation of ATP and NADPH, molecules that supply the chemical power needed to fix carbon dioxide into sugar (in a separate set of reactions that takes place outside the photosynthetic membrane). Photosynthetic bacteria have one photosystem of linked pigments and proteins. As shown here, cyanobacteria and green plants use two complete photosystems that work together; the chemical breakdown of water in Photosystem II provides the electrons needed for photosynthesis. (Reproduced with permission from W. K. Purves, D. Sadawa, G. H. Orians, and H. C. Heller, 2001. *Life: The Science of Biology*, Sixth Edition, Sinauer Associates and W. H. Freeman and Company)

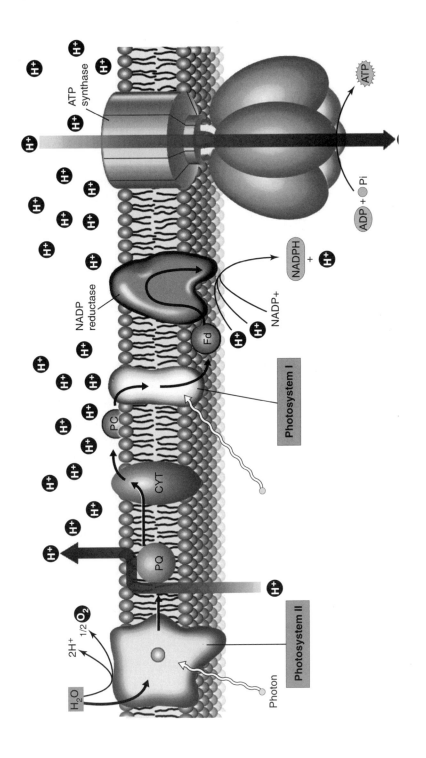

In metabolism, then, as in proteins, membranes, and nucleic acids, we can envision simple beginnings based on naturally occurring molecules, biological emancipation as biosynthetic pathways evolved, and, finally, complex biochemistry shaped by natural selection, gene duplication, and lateral transfer. The generative power of these processes is astonishing.

We are not close to solving the riddle of life's origins. Origin-of-life research resembles a maze with many entries, and we simply haven't traveled far enough down most routes to know which ones end in blind alleys. Yet, increasingly, chemists and molecular biologists have abandoned the early view that life originated by means of improbable reactions that came to pass only because vast intervals of time were available. Most now believe that life's origin (or origins—it could have happened more than once) involved chemistry that was both probable and efficient; there is a direct route through the maze, if only we can find it.

While we have no sharp constraints on the timetable of prebiotic evolution, it appears that by 3.8 billion years ago life may already have gained a beachhead on our planet. Some commentators worry that this leaves "only" a few hundred million years for life to emerge—but a hundred million years is a very long time! Asked to speculate on how long it took life to originate, Stanley Miller once suggested that "a decade is probably too short, and so is a century. But ten- or a hundred thousand years seems okay, and if you can't do it in a million years, you probably can't do it at all."

By 3.5 billion years ago, the metabolic diversification that ensured life's long-term perpetuation had almost certainly begun. Complex microbial communities cycled carbon and other elements through the biosphere. Even photosynthesis may have been present. Earth's oldest rocks, thus, lie near the juncture where the rootstock of primordial evolution meets the divergence of genes and organisms inferred from the Tree of Life.

Once life got going, where did it lead? By what route did the nascent biology of Warrawoona evolve over three billion years into the animals preserved in Kotuikan limestones? For that, we must resume our historical narrative.

6 | The Oxygen Revolution

Cherts of the Gunflint Formation, northwestern Ontario, pre-serve fossils of bacteria that lived nearly 2 billion years ago in an iron-rich sea. Even as the Gunflint rocks formed, however, Earth was completing a major environmental transformation—more than 2 billion years after our planet formed, oxygen began to pervade the atmosphere and surface oceans. The oxygen revolution redirected evolution, ushering in a new biological order that, far in the future, would lead to us.

"Take a good chunk. You never know when you'll be back." With that admonition, Elso Barghoorn heaved a fifty-pound block of chert into our aluminum dinghy, beached on a rocky promontory along the north shore of Lake Superior (figure 6.1).

It is 1974. For Elso, it is the autumn of a patriarch, a nostalgic return to the place where, twenty years earlier, he and Stanley Tyler changed the face of paleontology. For me, the trip provides a first opportunity to see the rocks that inspired me to follow in Elso's footsteps, a chance to contrast textbook certainty with outcrop reality, and an initiation into paleontology's oral tradition, passed on by the master as we walk along the shore. The rocks, strewn over our dented boat bottom and exposed in a thin bench along the shore, are Gunflint chert, at 1.9 billion years old nearly as distant in time from Warrawoona as we are from them.

Gunflint begins our return through time to the Cambrian. Do Gunflint fossils extend the enigma of Warrawoona forward or the familiar biology of Spitsbergen backward? As it turns out, they do a bit of both. But more than anything else, Gunflint cherts and associated iron tell

Figure 6.1. An outcrop of Gunflint chert along the north shore of Lake Superior. The figure in the distance is Elso Barghoorn, the father of Precambrian paleontology.

us that in middle age, Earth and life were going through important changes.

The Gunflint Formation is exposed along the lakeshore, in nearby road cuts, and in river-cut gorges of northwestern Ontario. Gunflint rocks include shales, a little bit of carbonate, and here and there some sandstone. A volcanic bed near the top of the formation contains zircons dated at 1,878 ± 2 million years. Most prominently, however, the Gunflint succession features iron formation, the remarkable rock of iron minerals and chert introduced in chapter 4. The banded iron formations of Warrawoona and the Barberton Mountain Land are among the oldest examples of this rock type; Gunflint's are some of the youngest. Iron, thus, provides our first clue to an Earth in transition.

The lowermost unit of the Gunflint Formation contains tiny fossils in fingerlike stromatolites of black chert. That sounds a lot like Spitsbergen, but the apparent similarity is deceptive. Examined closely, all aspects of Gunflint paleontology—fossils, stromatolites, and chert—differ from those of our younger example. In Spitsbergen and most other formations that contain silica-entombed microfossils, the chert occurs as lenses, nodules or thin beds that formed within carbonate sediments. Gunflint is different. These older cherts formed by silica precipitation directly on the seafloor.

Stromatolites reinforce the theme of difference. Gunflint structures (figure 6.2) may look broadly like the conventional stromatolites seen in limestones, but, in detail, many of them more closely resemble sinters—laminated buildups of silica precipitated from SiO_2-charged springs like those in Yellowstone Park. Sinters are the products of physical processes, although microorganisms can influence the details of lamination. The stromatolitic cherts of Gunflint may, therefore, reveal more about local chemistry than they do about microbial mats. In any event, Gunflint chert and iron tell us that this seaway was unusual—not like the tidal flats of Spitsbergen and, indeed, not all that similar to any habitat in the modern ocean.

Gunflint stromatolites contain huge populations of minute fossils, preserved along successive laminae within the buildups. Paleontologists debate whether these fossils record mat communities that built their stromatolitic mausoleums or bacteria that simply fell onto the accreting silica surfaces—like leaves entombed in Yellowstone sinters. My

Figure 6.2. Stromatolites in Gunflint chert. Each column is about 1 inch wide.

sympathies lie with the latter camp. The dense interweaving of filaments that identifies Spitsbergen populations as mat builders is rare in Gunflint stromatolites. Far more common are layers of microfossils jumbled together without consistent orientation, like flakes of parsley sprinkled on a casserole before serving.

The most common fossils are iron-coated tubes 1–2 microns across (plate 3a). Appropriately named *Gunflintia minuta*, these tiny threads resemble the cyanobacterial sheaths preserved in Spitsbergen cherts, but they also compare closely with the tubular sheaths of iron-loving bacteria such as *Sphaerotilus* and *Leptothrix* (plate 3b), found today wherever iron-rich waters come in contact with oxygen. Other fossils support the idea that iron-loving bacteria lived in the Gunflint sea. Spherical forms a few microns in diameter differ in detail from cyanobacterial cells, but resemble coccoidal iron-loving bacteria; twisted, branching tubes resemble the modern iron-lover *Gallionella*. Rarer fossils in Gunflint stromatolites include rod-shaped cells and colonies that could be the remains of cyanobacteria, but tiny starbursts (poetically named *Eoastrion*, the "little dawn star") strewn among the other fossils compare, once again, to bacteria that use iron and manganese in metabolism. Nonstromatolitic cherts deposited in quieter water contain starbursts in great abundance, along with small spherical cells that record some vanished plankton.

Like the fossils in Spitsbergen cherts, then, Gunflint microfossils resemble living microorganisms. But these older fossils are linked to a different set of modern counterparts—iron-metabolizing bacteria that are not much in evidence in today's iron-starved ocean. Microfossils, thus, confirm the environmental inference drawn from iron formations. The Gunflint ocean was not like that of Spitsbergen—or the seas we know today.

Sedimentary rocks 2.1–1.8 billion years old are widely distributed, and a dozen or so are known to contain fossils. Most come from iron-rich cherts, and they closely resemble the fossils in Gunflint rocks. Gunflint's iron-loving bacteria were not a local anomaly but rather a persistent feature of the global ocean.

Other discoveries, however, tell a different story. The most informative comes from the Belcher Islands, a cluster of low-lying islets near the eastern shore of Hudson Bay. Here, Montreal University paleontologist Hans Hofmann collected nodules of black chert in tidal-flat carbonate beds and found fossil cyanobacteria as strikingly modern as any in Spitsbergen (plates 3c and d). Iron-loving Gunflint microbes coexisted with mat-building blue-greens.

From this and other evidence, a fuller picture of early Proterozoic life begins to emerge. Stable isotopes suggest that two billion years ago, the microbial carbon and sulfur cycles operated much as they do today. Stromatolites are abundant in carbonate rocks of this age, and many can be attributed with confidence to the activities of microbial mats. Along with the small fossils in tidal-flat cherts, they contribute to a sense of early and long-lasting *Pax cyanobacteriana*.

Where iron-rich waters welled upward from the deep to mix with the oxygenated surface ocean, Gunflint-type bacteria flourished. Soon after Gunflint time, however, fossils of this type exited the geologic record. There is no reason to believe that Gunflint-type organisms were outcompeted by expanding cyanobacteria. After all, the two types of microbial community lived side by side for many millions of years. Rather, the paleontological demise of Gunflint-type assemblages reflects loss of habitat. By 1.8 billion years ago, iron formations—the lithologic signatures of iron-rich oceans—were gone.

As we ascend through the geologic column from Gunflint toward the Kotuikan cliffs, we can anticipate discovering fossils of increasing familiarity. But first, we must look backward and ask how life changed between Warrawoona and Gunflint times. As an exercise in micropaleontology, this is difficult, because fossils that might connect the two deposits are scarce. A single fossil in 2.7-billion-year-old cherts from Australia looks tantalizingly but not unambiguously cyanobacterial, and 2.5-billion-year-old cherts from South Africa contain indifferently preserved remains that could be blue-greens, as well. Fortunately, better evidence of cyanobacterial antiquity comes from an unexpected source—biomarker molecules in 2.7-billion-year-old shales found just south of North Pole in northwestern Australia.

The biomarker evidence was unexpected for two reasons. First, until recently, cyanobacteria were not known to produce molecular signatures that are both unique and preservable. But, of course, what organisms are "known to produce" can be quite different from what they actually make. Carefully integrating research in microbiology and organic chemistry, Roger Summons and his colleagues were able to identify distinctive lipid molecules that are synthesized in quantity only by blue-greens. Called 2-methylbacteriohopanepolyols (a name that is deeply meaningful to organic chemists and nearly incomprehensible to everyone else), these molecules are converted to molecules called 2-methylhopanes in sediments, where they can persist indefinitely as molecular fingerprints of ancient cyanobacteria.

Even accepting that cyanobacterial biomarkers can be recognized, however, their discovery in Archean shales was unanticipated. Biomarkers are destroyed by high temperatures, and conventional wisdom held that Archean sedimentary rocks were too "cooked" by metamorphism to yield well-preserved molecular fossils. Conventional wisdom isn't entirely wrong on this point—most Archean sedimentary rocks have turbulent geological histories that make them poor targets for molecular geochemistry. But not all Archean sediments have been roasted. The trick, then, is to ignore average rocks and concentrate on the few exceptional deposits that have avoided metamorphic alteration.

Working with Roger Summons and Roger Buick, Jochen Brocks, a Ph.D. student at the University of Sydney, did just that. In 2.7-billion-year-old shales that are both exceptionally well preserved and unusually rich in organic matter, Brocks found 2-methylhopanes, confirming the Archean origins of cyanobacteria. He found other biomarkers as well, including molecules called steranes. Steranes are geologically stable molecules derived from sterols, membrane-stiffening compounds made predominantly by eukaryotes.[1] (Cholesterol is the sterol best known to most of us.) Thus, by 2.7 billion years ago (or earlier!), the Tree of Life had begun to branch, producing diverse bacteria as well as the first buds on our own, eukaryotic limb of the tree (figure 6.3).

Our view of evolution between 3.5 and 1.9 billion years ago is biased by a sedimentary and paleontological record that increases in abundance and quality through time. We interpret this increase as a faithful chronicle of evolutionary diversification at our peril. Nonetheless, a

---

[1] The one group of bacteria known to synthesize cholesterol appears to do so using genes gained by lateral transfer from eukaryotes.

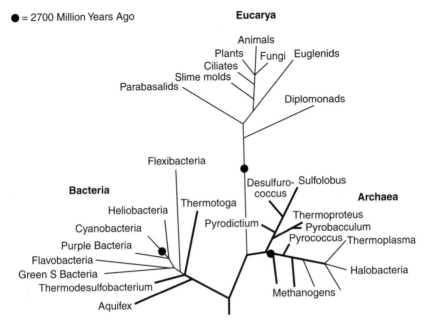

Figure 6.3. Molecular and isotopic biosignatures allow us to place time constraints on branch points in the Tree of Life. See text for discussion.

good argument can be made for dramatic biological changes not long before the Gunflint chert was deposited. The case, however, is not paleontological; it is geochemical.

In chapter 4, I argued that iron formations provide geological evidence for oxygen scarcity in the early Archean atmosphere and oceans. By extension, the persistence of iron formations until about 1.8 billion years ago suggests that the biosphere remained oxygen poor for a very long time (figure 6.4). Preston Cloud championed this idea four decades ago. Slight in stature, Pres was a giant of twentieth-century paleontology who recognized long before most others that biological and environmental history are intimately intertwined. He reasoned that if oxygen levels were low when life began but high today, we should look to the geological record for evidence of environmental transformation. For Cloud, and for many who followed in his footsteps, the stratigraphic distribution of iron formations (figure 6.5) focused the search on early Proterozoic rocks.

Figure 6.4. A mountain of iron. This landscape in Western Australia is carved out of a massive deposit of 2.5-billion-year-old iron formation.

Some of the most compelling evidence for oxygen scarcity on the early Earth comes from gravel and sand deposited by ancient rivers as they meandered across Archean and earliest Proterozoic coastal plains. Pyrite is common in organic-rich sediments, forming below the surface where H_2S produced by sulfate-reducing bacteria reacts with iron dissolved in oxygen-depleted groundwaters. Crystalline pyrite also occurs in igneous and hydrothermal deposits. Despite the fact that pyrite is common in rocks, however, it almost never contributes to the sediment grains formed when rocks erode. The reason is simple. Oxygen destroys pyrite, so on the modern Earth, pyrite disappears as it is exposed and eroded.

The same is true of two other oxygen-sensitive minerals: siderite (iron carbonate, or $FeCO_3$) and uraninite (uranium dioxide, or UO_2). Neither of these minerals is found today among the eroded grains that make up sediments on coastal floodplains, but both occur with pyrite grains in river deposits older than about 2.2 billion years. During the first half of Earth history, then, pyrite, siderite, and uraninite were exposed in rock faces, stripped away by weathering and erosion, and tumbled in rivers

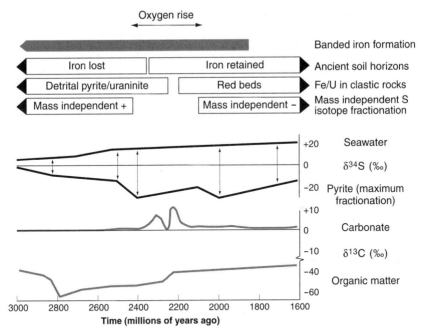

Figure 6.5. Summary of geologic evidence for environmental transition on the early Proterozoic Earth. See text for discussion.

until they came to rest in flood deposits—all without ever encountering oxygen in concentrations high enough to eliminate them.

As these oxygen-sensitive minerals faded from the scene, another oxygen-*requiring* rock type rose to prominence (figure 6.5). Visitors to Arizona or Utah take home vivid memories of canyons carved out of strikingly red sandstones and shales. These rocks—called red beds, in the button-down parlance of geologists—derive their color from tiny flecks of iron oxide that coat sand grains. The iron oxides form within surface sands, but only when the groundwaters that wash them contain oxygen. Red beds are common only in sedimentary successions deposited after about 2.2 billion years ago.

The simplest explanation for these observations is that prior to about 2.2 billion years ago, the amount of oxygen in the atmosphere and surface ocean was small. Once again, the question of how small still sparks debate, but if we avoid special pleading, the upper limit appears to be about 1 percent of present-day oxygen levels—and might have been much lower.

Independent evidence for early Proterozoic environmental change comes from ancient soils preserved by burial during floods. Soils form at the interface between rock and air, so they might be expected to reflect aspects of atmospheric chemistry. Dick Holland, a longtime friend and colleague at Harvard, has spent years hunting down ancient soil horizons and analyzing their chemistry. In fossil soils older than 2.4–2.2 billion years, he finds that the iron originally present in underlying rocks was removed as the soils formed. In contrast, iron in younger soils is retained (figure 6.5). Dick's explanation is that when parent rocks weathered under low oxygen conditions, iron was released as ferrous ions and carried away in solution by oxygen-poor groundwaters. In contrast, once oxygen became plentiful, iron released by weathering was immediately converted to insoluble iron oxides and, so, remained in place. Deriving quantitative estimates of atmospheric oxygen from these observations is a complicated business, requiring knowledge of parent rock chemistry and (poorly constrained) estimates of carbon dioxide levels in the ancient atmosphere. Dick's conclusion that atmospheric oxygen reached at least 15 percent of its present-day level may or may not be correct, but the *qualitative* conclusion that air became more breathable 2.4–2.2 billion years ago seems robust.

A loyal opposition, small but adamant, maintains that this record of atmospheric transformation is deeply misleading—that our oxygen-rich atmosphere originated much earlier than 2.2 billion years ago, perhaps even before Warrawoona time. Hiroshi Ohmoto, a geochemist at Pennsylvania State University and chief advocate of this alternative view, points out that mineralogical clues to ancient environments record *local* conditions that may not mirror the state of the planet as a whole. Ohmoto, therefore, interprets the evidence of iron formations, red beds, fossil soils, and O_2-sensitive minerals in terms of unusual Archean and earliest Proterozoic volcanic rocks, local oxygen depletion in marine basins, and the like. Ohmoto was particularly buoyed by Jochen Brocks's discovery of steranes in late Archean rocks, because sterol synthesis requires at least moderate amounts of oxygen (perhaps 1 percent of present-day levels, although the lower limit has not been established rigorously). Of course, what is good for the goose is good for the gander; the steranes might also record local rather than global oxygen abundance. Quite possibly—and consistent with the mineralogical evidence—sterol synthesis originated in local oxygen oases within cyanobacterial mats, only later to spread across the planet.

How do we adjudicate this debate? Is Earth's early sedimentary record really systematically misleading? Fortunately, some *biogeochemical* indicators provide globally integrated environmental signals, allowing us to evaluate mineralogical and biomarker data from a broader perspective. Principal among these are the isotopic abundances of carbon and sulfur in sedimentary rocks.

As explained in chapter 3, organic matter and limestones that accumulate on the present-day seafloor differ in their ratios of the stable carbon isotopes ^{13}C and ^{12}C by about 25 parts per thousand, reflecting the fractionation of carbon isotopes by photosynthetic algae and cyanobacteria. The isotopic differences between carbonate rocks and organic matter in most Precambrian sedimentary successions are only a little larger (26 to 30 parts per thousand)—the slight difference is thought to reflect similar biological processes played out beneath an atmosphere containing more carbon dioxide than at present. There are exceptions to this otherwise monotonous pattern, however, and, tellingly, almost all occur in rocks a bit older than 2.2–2.3 billion years.

In 1981, Martin Schoell and F. M. Wellmer discovered organic matter with unusually low ratios of ^{13}C to ^{12}C in lake beds about 2.8 billion years old from Canada. The organic matter was depleted in ^{13}C by as much as 45 parts per thousand, a fractionation too large to be ascribed to photosynthesis alone.

To understand these measurements and how they bear on Earth's oxygen history, we need to call on the Jacob Marley facts introduced in chapters 2 and 3. Earlier we learned that microorganisms have evolved diverse metabolisms and that some metabolic processes, notably photosynthesis, fractionate carbon isotopes as they work. Because photosynthetic (or chemosynthetic) organisms can't fractionate carbon isotopes by more than about 30 parts per thousand, we need to invoke additional metabolisms to explain Schoell and Wellmer's measurements. The most likely candidates are methane-eating bacteria at work in sediments. Methane eaters gain both carbon and energy from natural gas (CH_4), and, like photosynthetic organisms, they are choosy when it comes to isotopes. Because of their chemical preference for $^{12}CH_4$ over $^{13}CH_4$, microbes that eat methane fractionate carbon isotopes by 20–25 parts per thousand in environments where methane is abundant.

This allows us to account for the unusual chemical signatures in Schoell and Wellmer's lake beds. We begin with cyanobacteria that fractionate carbon isotopes by 30 parts per thousand, convert some of the organic matter they produce to methane, and then use this gas to feed hungry methane eaters that impart additional fractionation. The intermediate step is the trick. How do we convert cyanobacterial biomass into methane? Thinking back to chapter 2, the answer is methanogenic Archaea. Methanogens living in sediments gain carbon and energy by breaking down organic molecules to methane and carbon dioxide. When hydrogen is present, they can grow by chemosynthesis, as well, generating methane that is strongly depleted in ^{13}C. In combination, then, photosynthetic organisms, methane-producing archaeans, and methane-eating bacteria can explain the unusual isotopic values in late Archean lake deposits.

Methanogens play an important role in the carbon cycle of modern lakes. Knowing this, paleontologists believed that Schoell and Wellmer's discovery of high fractionation, 45 parts per thousand, made sense as a local, environmentally restricted exception to the rule. But it turned out not to be so exceptional. At about the same time that Schoell and Wellmer were working on Canadian rocks, John Hayes, an eminent geochemist now at the Woods Hole Oceanographic Institution, began a comprehensive survey of organic matter in Earth's oldest sediments. Hayes found carbon isotopic differences between carbonates and organic matter as large as 60 parts per thousand in late Archean and earliest Proterozoic rocks, and he found them in marine as well as lacustrine strata. Between 2.8 and 2.2 billion years ago, methanogenic Archaea must have enjoyed a prominence in the global carbon cycle that they have not commanded since that time.

If we wish to understand why methanogens were so important in early ecosystems, we must first ask what limits their abundance today. The reasons once again have to do with the varied forms of microbial metabolism introduced in chapter 2. In terms of energy yield, aerobic respiration is the favored pathway for breaking down organic molecules, so wherever oxygen is present, O_2-respiring organisms will dominate this leg of the carbon cycle. Within sediments, however, organisms use oxygen faster than it can be supplied from overlying waters. As a result, oxygen declines and, at some distance below the surface, disappears completely. (In lakes and coastal marine environments, oxygen can drop to zero within a few millimeters of the sediment surface.) Under these conditions, other metabolic pathways kick in. Nitrate respiration is next in line in terms of energy yield, but nitrate is generally in short supply, so these bacteria aren't major players in the carbon cycle. More important are sulfate-reducing bacteria. Sulfate is a major ion in seawater, enabling oxygen-depleted marine sediments to host large populations of sulfate reducers. Only where sulfate has been depleted, deep within marine sediments and at the bottom of the metabolic ladder, do we find fermenting bacteria and methanogenic archaeans. Lakes are a bit different. Because sulfate is only a minor constituent of fresh water, methanogens are more important than sulfate reducers in these settings.

We can now rephrase our question: why did the carbon cycle of late Archean and earliest Proterozoic oceans resemble that of modern oxygen-depleted lakes? Low oxygen levels provide an obvious explanation, or at least part of one. If oxygen was scarce on the early Earth, aerobic respiration must have been absent, or at least of limited and local biogeochemical importance. Oxygen alone doesn't solve the problem, however, since sulfate-reducing bacteria still dominate over methanogens in modern marine sediments. Perhaps sulfate, like oxygen, was scarce in early oceans.

Now we're closing in on our answer. Sulfate is produced in several ways. Photosynthetic bacteria can generate a limited supply, but most of the oceans' sulfate is formed when sulfurous volcanic gases combine with oxygen or when pyrite crystals react with oxygen during weathering. Thus, if oxygen was scarce on the early Earth, sulfate would have been, as well.

By calling once more on Jacob Marley facts from chapter 3, we can test the idea that Archean oceans were sulfate poor. Recall that sulfate-reducing bacteria fractionate sulfur isotopes much in the way that cyanobacteria fractionate carbon. Experiments on modern sulfate reducers show that the H_2S they produce can be depleted in ^{34}S by as much as 45 parts per thousand; however, where sulfate falls to levels below about 3 percent of present-day seawater, little isotopic fractionation takes place. Compilations by Donald Canfield, of Odense University in Denmark, show only limited isotopic fractionation in sedimentary sulfur from Archean deposits. Fractionation levels increase markedly in lower Proterozoic rocks, just as the exaggerated carbon isotopic signal associated with methane producers and methane eaters begins to tail off (figure 6.5). Isotopic measurements, thus, support the idea that oxygen levels rose early in the Proterozoic Eon, increasing the abundance of sulfate in seawater and, in consequence, reversing the importance of methanogenic archeans and sulfate reducing bacteria in the marine carbon cycle.

One more high-tech probe can be pressed into service. Sulfur comes in four isotopic varieties: ^{32}S, ^{33}S, ^{34}S, and ^{36}S. The ^{32}S and ^{34}S isotopes get most of the attention because they are abundant and easily measured. For most purposes we don't need to measure the rarer forms, because most processes that differentiate among isotopes do so by amounts that are directly proportional to their masses. Thus, if we know how fractionation has affected the abundant isotopes, we can calculate its effects on the rare ones.

I introduce this bit of chemical arcana because it leads us to an exciting new perspective on Earth's early environmental history. Although *most* chemical and biochemical processes fractionate isotopes in a mass-dependent fashion, a few—especially chemical reactions driven by light in the upper atmosphere—can partition isotopes in a way that is *independent* of their masses. Finding the chemical fingerprint of these processes in ancient rocks requires the painstaking measurement of sulfur in all its isotopic variety. Mark Thiemens and his team at the University of California, San Diego, figured out how to do just that. Their sensitive measurements of sulfur isotopes in samples of Mars delivered to Earth as meteorites showed that early in the history of our planetary neighbor, its sulfur cycle was dominated by atmospheric processes that imparted a mass-independent fractionation. In the wake of this discovery, James Farquhar, a postdoc in Thiemens's lab, trained his sights on ancient terrestrial rocks. To the great surprise of most geochemists, Farquhar demonstrated that gypsum and pyrite in Earth's oldest sedimentary successions *also* record mass-independent fractionation of sulfur isotopes. Like that of Mars, sulfur chemistry on the early Earth appears to have been influenced by photochemical processes that could be carried out only in an oxygen-poor atmosphere. Only after 2.45 billion years ago does this isotopic signal fade (figure 6.5), suggesting, independently of any other line of evidence, that oxygen began to accumulate in our atmosphere early in the Proterozoic Eon.

In summary, all biogeochemical signs point, as it were, to Rome. Around 2.4–2.2 billion years ago, it looks like the atmosphere changed. Hiroshi Ohmoto and his colleagues may have a point that oxygen began to accumulate earlier, perhaps locally and certainly in only trace abundances. But it was early in the Proterozoic Eon that the oxygenation of air and water assumed global environmental and biological importance.

Preston Cloud, Dick Holland, and other champions of early Proterozoic environmental transition were right. But why? What factors might push a planet from one long-lived environmental state in which oxygen was rare to another, where oxygen was relatively abundant? A simple answer might be that the evolution of cyanobacterial photosynthesis fomented early Proterozoic oxygen revolution. After all, photosynthesis is the principal source of O_2 on our planet. But the fossil record tells us that cyanobacteria began to diversify at least 300–500 million years before the atmosphere changed, and possibly much earlier.

To understand why photosynthesis alone could not sustain atmospheric transformation, you need only reflect that as you read these pages you are using atmospheric oxygen to respire organic matter, generating carbon dioxide and water in the process. Oxygenic photosynthesis and aerobic respiration form a tight couple in which the products of one metabolism provide the raw materials for the other. In a world where photosynthetic oxygen production is matched by respiratory oxygen consumption, O_2 cannot to build up in the atmosphere and oceans, regardless of how much photosynthesis takes place.

We need to envision processes that can break this coupling in ways that allow oxygen to accumulate. One possibility is to isolate organic matter from reaction with oxygen by burying it in sediments. This prospect changes the picture dramatically. What we previously thought of as a set of *biological* processes acquires a decidedly *geological* cast, because, globally, organic carbon burial is regulated by the dynamics of sedimentary basins and the deposits that fill them. Alternatively, we can decrease the rate at which oxygen is consumed by continental weathering and reaction with gases supplied by volcanoes. This, as well, inserts geology into Earth's carbon and oxygen cycles.

In practice, the photosynthesis/respiration couple is always a bit leaky, allowing a small fraction of the organic matter produced by photosynthesis to accumulate in sediments. Balancing this leakage, oxygen is always reacting with continental rocks and volcanic gases (often with the help of bacteria). To alter the face of the Earth, we need to look for *big* events.

Pres Cloud believed that iron in the Archean ocean sopped up the oxygen produced by early cyanobacteria, precluding O_2 buildup in the atmosphere. In his view, increasing photosynthesis by early Proterozoic cyanobacteria swept dissolved iron out of the deep oceans, releasing the brake on oxygen growth. This idea would be attractive, but for one untidy fact, already introduced. Iron formations didn't disappear 2.2–2.4 billion years ago, when other geologic indicators signal a rise in oxygen levels. The iron-rich rocks of Gunflint formed only 1.9 billion years ago, and a few other iron formations are still younger. This means that rusting oceans couldn't have released the oxygen brake. It also tells us that the O_2 generated 2.2–2.4 billion years ago was not sufficient to spread oxygen throughout the deep ocean.

Recently, David Catling, Kevin Zahnle, and Christopher McKay, all at NASA's Ames Research Center, have looked both above and below for explanations of early environmental evolution. They argue that on the late Archean and earliest Proterozoic Earth, some of the methane produced by all those methanogenic archaeans would have reached the upper atmosphere. There, ultraviolet radiation destroyed it, generating hydrogen gas in the process. Unlike most other gases, hydrogen is so light that it can break free of gravity and escape into space. Hydrogen loss would have made it easier for oxygen to gain a foothold at the Earth's surface. At the same time, rates of volcanism may have declined as the Earth's interior cooled, reducing the supply of oxygen-consuming gases to the atmosphere. Under these conditions, Catling and colleagues propose, oxygen began to accumulate in the atmosphere and surface ocean, building up until some other brake was applied.

As yet there is no consensus on the particular events that tipped Earth's environmental scale in planetary middle age. Whole-Earth models like that of Catling and colleagues are attractive, particularly because they are irreversible—once Earth went down the path of hydrogen escape, it could never recapture its oxygen-poor past. There is, however, one more line of evidence that requires consideration. And a third set of Jacob Marley facts.

Until now, our discussions of carbon isotopes have focused on the differences between carbonate rocks and organic matter. A much different type of information is encoded in the *absolute* values of $^{13}C/^{12}C$ in these materials. As shown in figure 6.6, the higher the $^{13}C/^{12}C$ values of carbonates and organic carbon, the higher the rate of organic matter burial (relative to carbonate deposition) at the time the sediments formed. Carbon isotopic values of early Proterozoic limestones and dolomites are the highest ever recorded globally, supporting the hypothesis that geological changes contributed to the oxygen revolution by promoting the burial of organic matter in sediments.

David Des Marais, also at NASA Ames, has calculated that the amount of oxygen generated by excess organic carbon burial 2.4 to 2.2 billion years ago would have been enough to generate present-day oxygen levels ten times over. The persistence of iron formations until 1.85 billion years ago, however, shows that it didn't do so. Where did all the oxygen go? Most of it combined with sulfur to form sulfate, giving the sea its modern tang.

Don Canfield was the first to point out an important consequence of this change. We noted earlier that as sulfate levels rose in the ocean, sulfate-reducing bacteria increased concomitantly in importance. Hydrogen sulfide is the by-product of sulfate reduction, so as populations of sulfate-reducing bacteria swelled, more and more H_2S would have been produced in the deep sea. H_2S reacts readily with dissolved iron to form pyrite, providing an alternative explanation for the loss of iron formations. Hydrogen sulfide, and not oxygen, might have swept iron from the sea, leaving the deep ocean as anoxic as it was when Warrawoona was young.

How did biology respond to the oxygen revolution? We read, provocatively, of an "oxygen holocaust" in which untold lineages of anaerobic

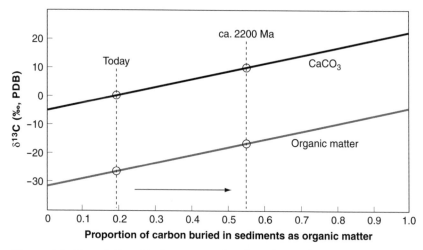

Figure 6.6. The relationship between carbon burial in sediments and the isotopic composition of carbonates and organic matter, after a diagram by John Hayes. Carbon entering the Earth surface system from the mantle (by way of volcanoes) has a $\delta^{13}C$ value of about –6‰. (The "delta" notation used by geochemists indicates the difference between $^{13}C/^{12}C$ in the sample and that of a laboratory standard, expressed in parts per thousand—symbolized by ‰.) If all carbon entering the system were deposited as carbonate, the $\delta^{13}C$ value of that carbonate would also be –6‰, because, in terms of isotopes, what comes out must equal what went in. For the same reason, if all carbon were buried as organic matter, *its* $\delta^{13}C$ value would be –6‰. In the real world, where carbon enters sediments as a mixture of carbonate and organic matter, the total isotopic composition of carbon leaving the system must still match that coming in; this is achieved when isotopic compositions for carbonate and organic matter follow the diagonal trend lines shown in the graph. Today, for example, carbon burial in sediments is about 81% carbonate and 19% organic matter, and $\delta^{13}C$ values of carbonate and organic matter are about 0‰ and –28‰, respectively. In 2.2-billion-year-old rocks, however, carbonate $\delta^{13}C$ values are commonly about +8‰, whereas the $\delta^{13}C$ values of organic matter hover around –20‰, suggesting that during this interval, rates of organic carbon burial matched those of carbonate deposition.

microorganisms perished. But anoxic environments didn't disappear 2.2 billion years ago; they simply retreated beneath an oxygenated veneer of surface sediments and water. Indeed, rather than considering the early Proterozoic as a time of environmental *transition*, it may be more profitable to think of it as an interval of environmental expansion—one that that enabled the Earth to support an unprecedented diversity of life. Anaerobic microorganisms retained their critical roles in ecosystem function, roles that they retain today. But organisms that use or at least tolerate oxygen expanded greatly. Aerobic respiration became a dominant metabolism among bacteria, and chemosynthetic bacteria that gain energy by reacting oxygen with hydrogen or metal ions diversified along the interface between oxygen-rich and oxygen-depleted environments.

In Gunflint time, halfway through our planet's history, the Earth remained an unfamiliar place. But the trajectory of subsequent evolution had been set. From this time onward, organisms that use or produce oxygen would dominate biology. Indeed, at the Earth's surface, *only* oxygen and carbon dioxide would ever again be abundant enough to supply the needs of cells larger than a few microns, and oxygen would eventually achieve concentrations able to support large, multicellular organisms. From now on, the Earth would start to become *our* world.

7 | The Cyanobacteria, Life's Microbial Heroes

If oxygen wrought revolutionary change, cyanobacteria were the heroes of the revolution. Exceptionally preserved fossils in 1.5-billion-year-old cherts from Siberia show that blue-greens diversified early and continue today in little-altered form. The capacity to change rapidly but persist indefinitely may epitomize bacterial evolution.

THE ASCENT UP the Great Wall is tiring. It is cold and it's damp, there are precious few places to rest, and the footfalls can be slippery. Thank goodness there are no other visitors.

No other visitors? Veterans of Beijing tourist itinerary A (Great Wall in the morning and Summer Palace in the afternoon, with factory tours discontinued) may startle at the thought, but like my North Pole, this isn't the Great Wall of common experience. My Great Wall is an aptly named sliver of dolomite miles long and hundreds of feet high that separates two pristine rivers in northern Siberia (figure 7.1). On the north flank, a familiar stream carries its load of silt and snowmelt westward toward the Arctic Ocean. It is the Kotuikan, the same watery ribbon that carved the Cambrian cliffs encountered in chapter 1. The dolomites have also been introduced before. They're part of the thick sedimentary pile glimpsed downstream beneath the Kotuikan's Proterozoic-Cambrian boundary succession. In the Great Wall, these older rocks are breathtakingly well exposed—a slender mesa of carbonate beds that lie as flat today as they did when they formed nearly 1.5 billion years ago.

Figure 7.1. The Great Wall along the Kotuikan River in northern Siberia. The wall is built of flat-lying carbonate rocks deposited along the edge of the ocean some 1.5 billion years ago.

Volodya Sergeev, resplendent in army jacket and Red Sox baseball cap, has brought me here on a drizzly day in late June. Despite the weather, there is uncommon pleasure in climbing this remote ridge. Below, the taiga is shedding its winter gray for the vibrant hues of a short but intense summer. Early leaves tint the larches grass-green; roses and peonies seem to bloom everywhere; even the local foxes are shedding winter coats for summer. Owls perched on overhead branches distract us, but mosquitoes don't; mercifully, we're still a week away from their annual convocation. As Volodya and I pick our way up the cliff face, we debate good naturedly about the rocks beneath our feet. Each bed is scrutinized carefully. If it won't support our arguments, it must at least support our weight.

The Bil'yakh Group—the formal name given to the Great Wall dolomites and associated rocks—continues our ascent through time. The leap from Gunflint is substantial, covering nearly a third of the 1.35-billion-year distance between those iron-rich cherts and the Cambrian. Having made the jump, what do we see? Cyanobacteria everywhere.

The Great Wall is a good place to focus on blue-greens, arguably the most important organisms ever to appear on Earth. Cyanobacterial fossils seen earlier in cherts from Spitsbergen and the Belcher Islands hint at a general, and remarkable, feature of this group—populations preserved 750 million, 1 billion, or even 2 billion years ago are essentially indistinguishable from living forms. This is quite different from the fossil records of plants and animals, which are replete with extinct forms. Why should the evolutionary history of cyanobacteria, so much longer than that of animals, be more static, as well? This fundamental question, framed by fossils, deserves a thoughtful answer. We'll be in a better position to attempt one after considering the illuminating treasures in Great Wall chert.

Chert nodules in the Bil'yakh Group contain lots of cyanobacterial fossils, many of them slender filaments frozen in tufts that stand tiny but erect in the rocks (plate 4a). More abundant, however, are spheroidal cells packed into globular colonies, like tiny cumulus clouds, along bedding surfaces (plate 4c). *Eoentophysalis* is the modern-looking cyanobacterium discovered by Hans Hofmann in 2.0-billion-year-old cherts of the Belcher Islands, and here, in northern Siberia, it built equally extensive mats 1.5 billion years ago. These distinctive microfossils have cells that look and divide like modern *Entophysalis* (plate 3d), arrayed into pigmented colonies like modern *Entophysalis*, which connect to form mats like modern *Entophysalis*, in environments like those in which *Entophysalis* thrives today. You get the picture.

Other fossils include short filaments (plate 4b) and tiny rods that divide by binary fission; however, Spitsbergen favorites like the stalk-forming *Polybessurus* are nowhere to be seen.

One particular Bil'yakh population deserves a closer look. *Archaeoellipsoides* is a big (in relative terms!) sausage-shaped microfossil first discovered in northern Canada by the late Bob Horodyski (plate 4d). These fossils are common in mid-Proterozoic cherts, but their biological interpretation remained elusive until specimens from the Great Wall provided an ID. Specifically, the sizes (up to 100 microns long) and shapes of the Siberian fossils, their lack of evidence for cell division, and the presence of germinating filaments in close proximity combined to mark *Archaeoellipsoides* as the reproductive spores of a filament-forming cyanobacterium. Modern *Anabaena* makes comparable structures.

That's good—another fossil cyanobacterium with a close living counterpart. There is, however, special reason to be interested in *Archaeoellipsoides*.

The differentiation of specialized cell types is commonplace in animals, but rare in cyanobacteria. Only a few blue-greens are able to accomplish this trick, and they all fall on a shallow branch of the
cyanobacterial tree (figure 7.2). Thus, if we see *Archaeoellipsoides* in 1.5-
billion-year-old rocks, the evolutionary diversification of cyanobacteria
inferred from molecular phylogenies must have taken place earlier. Indeed, we can push a bit further back—Janine Bertrand-Sarfati, of the
University of Montpellier in France, has identified *Archaeoellipsoides* in
West African cherts nearly 2.1 billion years old.

The modern counterparts of *Archaeoellipsoides* actually differentiate
two specialized types of cells. In addition to reproductive spores that
can enter the fossil record as identifiable structures, these blue-greens
form thick-walled (but not easily preserved) cells that are specialized for
nitrogen fixation. Nitrogen fixation is extremely sensitive to oxygen—
even modest concentrations of O_2 inhibit this process. The specialized
cells in *Anabaena* and related cyanobacteria keep oxygen from diffusing
into their interiors, thereby providing localized sites for nitrogen fixation
in an oxygen-rich world.

The oxygen revolution discussed in chapter 6 would, thus, have provided environmental impetus for the evolution of specialized cells in
blue-greens. As *Archaeoellipsoides* fossils appear in the geological record
by 2.1 billion years ago, we have a reasonable estimate of when the main
branches of the cyanobacterial tree were established. By the time that
microfossils first bring them into sharp focus, cyanobacteria must already have evolved much of their present-day diversity.

Paleontologists love to read stratigraphic pattern as evolutionary history, and the history suggested by Proterozoic fossils is that cyanobacteria evolved early and quickly, and then just sat there, changing little
over the eons. Alternatively, however, we might imagine that while the
simple shapes of cyanobacteria have remained constant through Earth
history, physiology has not—something Bill Schopf long ago (but with
recently restored resonance) termed the Volkswagen Syndrome. As
noted in chapter 3, there is reason to be skeptical about this reading,

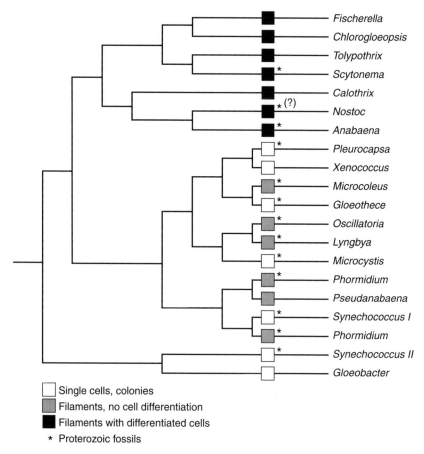

Single cells, colonies
Filaments, no cell differentiation
Filaments with differentiated cells
* Proterozoic fossils

Figure 7.2. A tree showing evolutionary relationships among living cyanobacteria. Note that cyanobacteria with specialized cells fall on a fairly late branch of the tree. This means that fossils showing cell differentiation can place an upper bound on when the tree's major branches formed. (Phylogenetic data courtesy of Akiko Tomitani)

because the similarities between ancient and modern cyanobacteria extend beyond form to include physiologically determined features such as life history, behavior, and environmental tolerance. Also, many features of cyanobacterial biology are conserved across the entire phylum and so must already have been present when blue-greens began to diversify.

A more subtle problem concerns convergence. Perhaps the apparent pattern of long-term persistence in specific habitats reflects instead the repeated molding of form and physiology by environment. If this were true, then the 2-billion-year record of *(Eo)Entophysalis* would tell us only that whenever arid tidal flats became established, cyanobacterial colonists evolved features like those of modern entophysalids. The convergence argument is hard to shake using fossils alone, but comparative biology provides reasons for doubt. If convergence rather than evolutionary relatedness explains the fit between form and environment, then we might expect morphologically similar cyanobacteria to crop up on different branches of phylogenetic trees constructed from molecular data. In fact, simple unicellular and filamentous forms do appear to have evolved repeatedly within the blue-greens—score one for convergence. In contrast, morphologically complex cyanobacteria—the forms highlighted in this book because they cannot be mistaken for other sorts of bacteria—cluster together as coherent branches in the cyanobacterial tree. For these taxa, similarity of form reflects common ancestry.

In the end, despite questions of physiology and convergence, the simplest reading of the early fossil record is probably the best one. Cyanobacteria arose long ago and early on evolved most of the molecular and morphological features seen in their living descendants.[1]

Now, at last, we can return to the central riddle of cyanobacterial evolution, confident that we are framing the question correctly. Why have many blue-greens persisted so long with so little change?

Long-term stasis occurs because populations neither die off nor change. That may seem trivially obvious, but it emphasizes the fact that we have two features to explain. The general resistance of bacteria to extinction is well known. Bacteria have immense population sizes, and they can reproduce rapidly—it doesn't matter how thoroughly you clean your teeth in the morning; the bacteria that evade your toothbrush will multiply to film your mouth by late afternoon. Bacteria also track shifting environments with ease. The air, for example, is full of bacteria,

---

[1] This does not mean that Great Wall cyanobacteria were identical in molecular detail to their modern counterparts. Without doubt, the nucleotide sequences for many genes have changed somewhat through time, and some enzymes in living populations are more efficient than those in long-dead ancestors. What I *do* mean to convey is that living cyanobacteria provide pretty specific guides to the functional biology of fossils preserved in Great Wall cherts.

and if you place a bowl of milk on a windowsill, it will be cheese before long. Further, bacteria are good at withstanding environmental disturbance. Although most bacterial strains grow best within a narrow habitat range, they can tolerate much more extreme conditions, at least for a short time.

Bacteria are particularly good at doing nothing. When the surrounding environment is favorable for growth, bacteria multiply rapidly, as they do in your mouth. But when ambient conditions do not favor growth, they are able to persist in a dormant stage, with little expenditure of energy. Actually, most bacteria at most times may exist in a state of metabolic torpor, ready to spring into action the moment that resources become available.

Such features explain why bacteria in general, and cyanobacteria in particular, should be persistent. But why don't they change? Why should fossils from a 1.5-billion-year-old tidal flat look just like the cells observed in coastal mats today? The paleontological observation of long-term cyanobacterial stasis is particularly puzzling because we know that bacteria *can* evolve rapidly. New, disease-resistant strains of rice and wheat last only a decade or so before some microbe learns to take their measure. Evolved bacterial resistance to antibiotics looms as a major issue in public health.

In the laboratory, we can inoculate a culture medium with bacteria that do not grow naturally on the nutrients contained in the medium. Most cells won't grow in their new surroundings, but a few harbor mutations that allow them to use the novel nutrients. Initially, the mutants fare poorly, but even all-thumbs growth allows survival. As mutations continue through time, natural selection hones metabolic efficiency on the new substrate. Adaptation to novel environments can easily be investigated in the course of a Ph.D. thesis or federal grant—critical timescales for laboratory biologists, but instantaneous by geological standards.

The apparent tension between rapid evolution and prolonged stasis can be resolved with help from an evolutionary metaphor introduced by Sewall Wright in 1932. In any given environment, some combinations of genes serve their owners better than others. Through time, then, natural selection will favor genetic types that grow and reproduce best. Wright thought about this interplay between genes and environment in terms of an *adaptive landscape*. Each point on the metaphoric landscape represents a specific combination of genes; topography indicates how well each combination works in the environment. (In evolution-speak, the points are *genotypes* and the hills and valleys indicators of *fitness*.) New populations migrate toward higher ground, driven by natural selection to surmount a peak of fitness. It needn't be the highest point on the landscape—commonly it will be the local hill most easily approached from the population's starting point—but gene combinations that form the valley floor don't survive for long.

Karl Niklas, a botanist at Cornell University, used a simple mathematical model to explore why some adaptive landscapes are rugged and others gentle. He found that when organisms must do many things at once, trade-offs in form and physiology enable many different genetic combinations work about equally well—the adaptive landscape consists of low rolling hills, like the Cotswolds or Pennsylvania Dutch farmland. Adaptive landscapes probably look like this for most plants and animals. In contrast, when only one functional demand must be satisfied, a single peak dominates the adaptive landscape. Bacteria can be famously single-minded.

Experiments by Richard Lenski, a microbiologist at Michigan State University, lend credence to this view. Lenski introduced a population of *E. coli* into a novel culture medium and then followed the population day by day for 10,000 generations (not quite five years). For the first 2,000 or so generations, successive generations got better and better at living in their lab environment. After that, however, improvement slowed and eventually stopped. The population evidently reached a point where additional mutations had a vanishingly small probability of improving performance.

Niklas's models and Lenski's experiments help to reconcile biological evidence for rapid bacterial evolution with paleontological observations of cyanobacterial stasis. For cyanobacteria, adaptive landscapes probably resemble Mt. Fuji rising out of the plains of Kansai. Early in Earth history, newly evolved blue-greens invaded tidal flats and other environments. In each new habitat, natural selection pushed cyanobacterial populations up a steep adaptive peak. Those that reached the summit were difficult to dislodge and unlikely to descend. If this picture is even approximately correct, cyanobacteria should adapt rapidly (on a geological timescale) to new environments and then persist as long as the environment is present. That, of course, is just what we see in the Proterozoic fossil record.

More generally, this view suggests that on the timescale of Earth history, the tempo of bacterial evolution is determined by rates of environmental change. New habitats beget new adaptations, with the result that bacterial diversity has expanded as the range of habitable environments has grown. The evolutionary processes at work are Darwinian, and the pattern that results is reminiscent of Eldredge and Gould's punctuated equilibrium—except that extinction is rare and diversity, therefore, accumulates. Of course, environments can be biological as well as physical; evolving plants and animals have simply provided the bacteria with new kingdoms to conquer.

The Great Wall fossils shed light on another riddle of Proterozoic rocks, this one posed by stromatolites, the laminated structures found in so many Proterozoic limestones and dolomites. In the 1950s, Russian geologists began the herculean task of mapping Precambrian rocks across the vast Siberian platform. Thick piles of Proterozoic sediments occur in many parts of this gargantuan landmass, mostly hidden beneath forests and swamps, but here and there excavated by rivers like the Kotuikan. Mapping requires an understanding of how scattered outcrops relate to one another, but the Proterozoic beds of Siberia don't contain shelly fossils, the conventional guides to geological correlation. On the other hand, they are chockablock with stromatolites. Along the Kotuikan River, thick stromatolitic reefs can be followed for miles (figure 7.3), and similar features occur in Proterozoic carbonates throughout Siberia. Some of the stromatolites are club-shaped, others conical; some branch regularly, others not at all (figure 7.4). Details of lamination and microscopic texture vary, as well.

Russian geologists—including Misha Semikhatov, whom we met in chapter 1—described these features in exacting detail and, in the process, became convinced that stromatolites held an important key to the correlation of Proterozoic rocks. They were right. Middle and late Proterozoic stromatolites can easily be distinguished from one another, and early Proterozoic stromatolites are something else again. What's more, the stratigraphic pattern seen in Siberia and the neigh-

Figure 7.3. Stromatolitic reefs in the 1.5-billion-year-old Bil'yakh Group, northern Siberia. The bunlike feature to the right of Misha Semikhatov (who, for scale, is precisely 2 meters tall when wearing a hat) is a small reef. Misha is standing on the curved upper surface of a second, larger reef. And the wall extending above him is part of still another reef, this one the size of a small office building.

boring Ural Mountains is reprised in Proterozoic rocks around the world.

Therein lies what Shakespeare called the rub. Stromatolites are built by cyanobacteria, but as we've already seen, cyanobacteria show little evidence of an evolutionary trajectory through the long Proterozoic Eon. Why, then, should stromatolites change through time?

Bil'yakh fossils suggest an answer. Dolomites in the Great Wall contain sedimentary features we've seen before: crinkly laminated mats, tepee structures, sheets of ooids, and low stromatolitic domes. Like the Spitsbergen rocks introduced in chapter 3, Great Wall beds accumulated along the edge of an ancient ocean, in tidal flats and adjacent coastal environments. Despite similarities in habitat, however, the fossils in Bil'yakh cherts differ substantially from those in the Akademikerbreen Group. *Eoentophysalis*, so abundant in the Great Wall, is rare in the younger Spitsbergen rocks. Conversely, the stalked remains of *Poly-bessurus* found in Spitsbergen cherts do not occur in the Great Wall

Figure 7.4. Stromatolites in mid-Proterozoic carbonate rocks from Siberia. The largest column is about 4 inches wide.

assemblage. We can't blame the differences on evolutionary change; the dominant cyanobacteria in Bil'yakh and Akademikerbreen cherts *all* have close counterparts among living blue-greens.

The alternative is that the *environment* changed through time. The vertical tufts of blue-green filaments preserved in Great Wall cherts provide subtle but important support for this idea. Tufts are common in modern mats, but they don't persist as recognizable fabrics in sediments, because the weight of overlying deposits flattens them. The Bil'yakh filaments remain vertical because they were stiffened by calcium carbonate cement *before* they were buried. We didn't observe this in Spitsbergen. It suggests that the Great Wall tidal flat was a bit different from its younger counterparts, more prone to carbonate precipitation right at or just beneath mat surfaces.

Other observations support this conclusion. For example, in modern tidal-flat sediments (and in Spitsbergen cherts), accumulating sediments collapse buried cells, compressing them as they decay. On the Great Wall tidal flat, however, decaying cells left behind three-dimensional voids now filled by cement—when these cells decayed, surrounding sediments had already become rock (plate 4b). Research by Julie Bartley,

Plate 1. Microbial ecosystem around a hot spring at Orakei Korako, New Zealand. The long blue-green streamers consist of cyanobacteria. Populated by metabolically diverse Bacteria and Archaea, modern hot springs suggest what life may have been like on the early Earth.

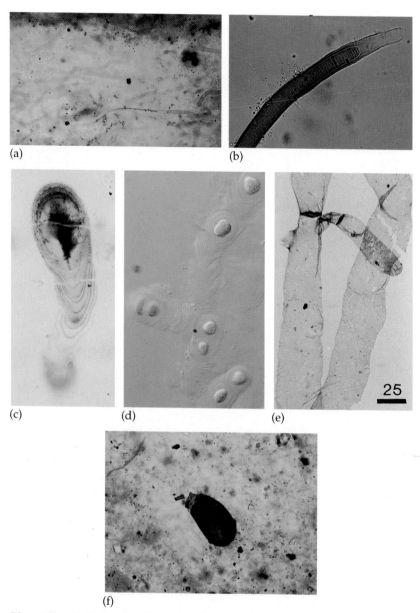

(a)

(b)

(c)

(d)

(e)

25

(f)

Plate 2. Fossils in Akademikerbeen cherts and shales, and some living counterparts. (a) Filamentous fossils of mat-forming microorganisms in Spitsbergen chert; each tube is about 10 microns wide. (b) The cyanobacterial genus, *Lyngbya*, which provides a modern counterpart for the fossils in a. (The specimen is 15 microns wide.) Note the extracellular sheath that surrounds the ribbon of cells. Because it is not easily destroyed by bacteria, this sheath, rather than the cells it contains, is likely to enter the fossil record. (c) *Polybessurus bipartitus*, a distinctive stalk-forming microorganism in Spitsbergen cherts; specimen

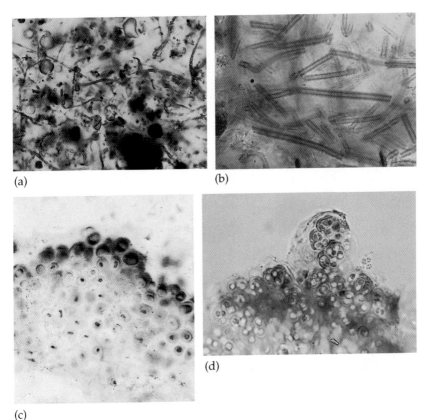

(a)

(b)

(c)

(d)

Plate 3. Early Proterozoic microfossils and their modern counterparts. (a) A microscopic view of Gunflint chert, chockablock with tiny fossils. (b) *Leptothrix*, a modern iron-loving bacterium thought to be similar to the filaments in Gunflint fossil assemblages. In both figures, the filamentous organisms are 1–2 microns across. (c) *Eoentophysalis* cyanobacteria in early Proterozoic chert from the Belcher Islands, Canada. (d) A modern *Entophysalis* species for comparison (ellipsoidal envelopes around cells are 6–10 microns wide in both illustrations). (Photo (c) courtesy of Hans Hofmann; photo (d) courtesy of John Bauld)

Plate 2 *(continued)* is about 35 microns wide. (d) A modern stalk-forming cyanobacterium that forms crusts on tidal flats of Andros Island, Bahamas; each specimen is 15 microns wide. This living counterpart to *Polybessurus* was discovered on the basis of environmental predictions made from the 600–800-million-year-old fossils. (e) Multicellular fossil from Spitsbergen shale comparable to the living green alga *Cladophora* (tubes 25 microns across). (f) Vase-shaped microfossils of tiny protozoans (fossil is 100 microns long; see chapter 9 for further discussion). (Image (b) courtesy of John Bauld; (e) courtesy of Nicholas Butterfield)

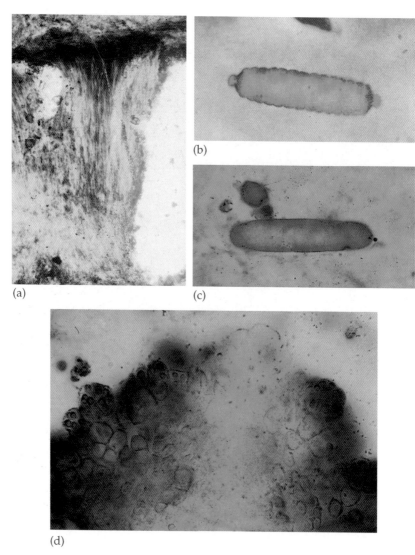

Plate 4. Cyanobacterial microfossils in cherts of the 1.5-billion-year-old Bil'yakh Group. (a) A vertical tuft of tubular filaments, preserved in this orientation by very early formation of calcium carbonate cement (each filament is about 8 microns across). (b) A filamentous cyanobacterium, showing how cells were arranged along its length; the specimen is actually preserved as a lightly pigmented cast, originally made in rapidly cemented carbonate sediment (fossil is 85 microns long).(c) *Archaeoellipsoides*, the large (80 microns long, in this case) cigar-shaped fossil interpreted as the specialized reproductive cell of an *Anabaena*-like cyanobacterium. (d) 1.5-billion-year-old mat-building colony of *Eoentophysalis*; see plate 3d for its modern counterpart.

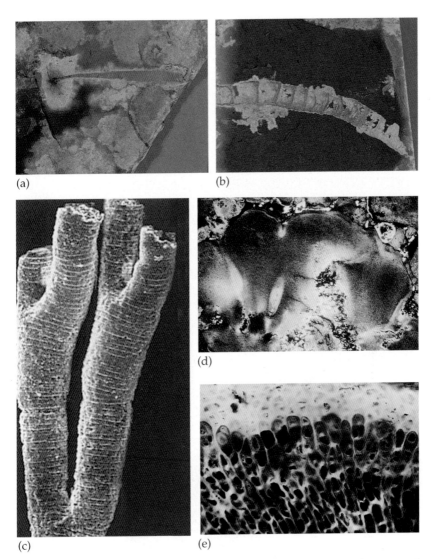

(a)

(b)

(c)

(d)

(e)

Plate 5. Eukaryotic fossils in Doushantuo rocks. (a) Compression of a seaweed from shales at Miaohe; specimen is 2 inches long. (b) Tubular fossils of unknown but possibly animal origin, also from Miaohe; specimen is 3 inches long. (c) Small (150 microns across) branching tubes with distinctive cross-walls preserved in Doushantuo phosphates; possibly, these were made by early relatives of corals. (d) and (e), Multicellular red algae in Doushantuo phosphates; (d) illustrates a section through an alga, showing "cell fountains" and cylindrical recesses interpreted as reproductive structures; specimen is 1 millimeter across. Dark spots are individual cells; (e) shows well-preserved cells (6–10 microns in diameter) at higher magnification.

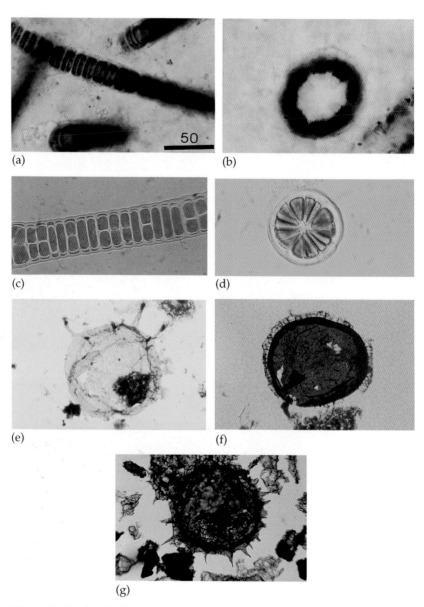

Plate 6. Fossils of Proterozoic eukaryotes. (a) and (b) illustrate the fossil *Bangiomorpha* in ca. 1.2-billion-year-old cherts from arctic Canada. (c) and (d) show the living red alga *Bangia*. All specimens are about 60 microns in cross-sectional diameter. (e) *Tappania*, a 1.5-billion-year-old microfossil from northern Australia; fossil is 120 microns wide. (f) A lavishly ornamented microfossil (200 microns in diameter) interpreted as the reproductive spores of algae from ca. 1.3-billion-year-old rocks in China. (g) A large (more than 200 microns) spiny microfossil from ca. 570–590-million-year-old rocks in Australia. (Photos (a)–(d) courtesy of Nicholas Butterfield)

(a)

(b)

(c)

(d)

(e)

Plate 7. Ediacaran fossils from Namibia and elsewhere. (a) *Swartpuntia*, a three-winged fossil found in the uppermost Proterozoic beds of the Nama Group; only two "wings" are evident in these fossils. (b) *Mawsonites*, a 4-inch disk from South Australia, interpreted an a sea anemone–like animal or the holdfast of a sea pen–like colony. (c) *Dickinsonia*, the most celebrated (and controversial) of vendobiont fossils. This specimen is from the Ediacara Hills of South Australia. (d) *Beltanelliformis*, a spherical green alga, here seen in latest Proterozoic sandstones from the Ukraine; specimens 1/2 to 3/4 inch across. (e) *Pteridinium*, another three-winged fossil found in sandstones of the Nama Group. (Photos (b) and (c) courtesy of Richard Jenkins)

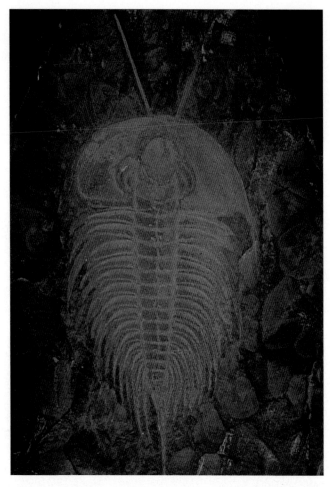

Plate 8. The trilobite *Olenellus*, illustrating the tremendous complexity achieved by Early Cambrian animals. (Photo courtesy of Bruce Lieberman)

an alumna of my lab who is now at West Georgia University, indicates that the timescale for cyanobacterial decay is commonly days to weeks—those Great Wall carbonates hardened fast. A few beds in the Bil'yakh succession even display the finely layered seafloor precipitates that complicate our interpretation of stromatolites in much older rocks. Along the Great Wall coastline, then, carbonates accumulated like papier-mâché, entombing microorganisms and providing a distinctively firm seafloor for colonization. Stromatolite accretion reflects both life and environment. Thus, changing seawater chemistry helps us to understand why stromatolite forms vary through time.

In the early 1990s, I had the good fortune to study an invaluable collection of thin sections prepared from Siberian stromatolites by the late V. A. Komar, one of the resourceful geologists who pieced together the Proterozoic history of Siberia. Along with Misha Semikhatov, I spent many hours trying to understand the paleobiological messages encrypted in the microscopic fabrics of these rocks. We found that stromatolites built by microbial trapping and binding of fine-grained sediments generally exhibit the uninspiring (and uninformative) texture of mud—layered, uniform, and bland. In some samples, however, stromatolites formed principally by calcium carbonate precipitation contain carbonate-encrusted filaments that lie in tangles along laminae. Evidently, filamentous sheaths provided microscopic sites for the precipitation of fine calcium carbonate crystals.

Interestingly, encrusted filaments occur only in stromatolites younger than about 1.0 billion years. Mid-Proterozoic stromatolites formed by carbonate precipitation also display distinctive microfabrics, in this case vertical splays of crystals formed at or just beneath the seafloor. Going backward from late to middle Proterozoic, then, the fine calligraphy of encrusted filaments gives way to a coarser crystal-dominated texture that masks the signature of mat-building microorganisms. Of course, as we retreat still further back in time, we encounter greater volumes of *macroscopic* carbonate precipitates made of stacked crystal fans (figure 7.5). Stromatolite fabrics, thus, reprise a theme developed earlier: the older the carbonate, the greater the evidence for cement precipitation at or near the seafloor surface. Proterozoic stromatolites appear to be environmental dipsticks, recording episodic changes in the carbonate chemistry of seawater as atmospheric CO_2 levels declined and O_2

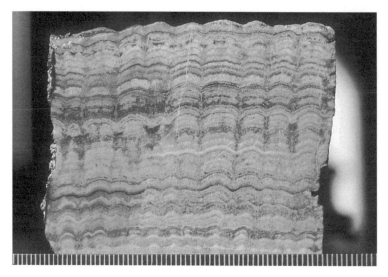

Figure 7.5. Fingerlike laminated structures, each 0.4 inches across, formed by calcium carbonate precipitation without any obvious participation by microbial mats. This specimen, collected by Linda Kah, comes from a 1.2-billion-year-old tidal-flat deposit on Baffin Island, northern Canada.

increased. Change through time in stromatolite form is fully consistent with evolutionary stasis in cyanobacterial mat builders.

To the extent that evolution has influenced stromatolite history, it may have done so primarily in a rather different way. As seaweeds diversified in the late Proterozoic Eon, they began to form algal lawns where microbial mats once held sway. The subsequent diversification of animals further intensified competition for space on the seafloor, and introduced grazing as well. In consequence, for the past 500 million years, stromatolites have largely been limited to lakes and restricted coastlines where competition and predation are strongly limited—reclaiming something of their former glory only in the aftermath of mass extinctions, and then only transiently.

Returning one last time to the shores of the Kotuikan River, we can contemplate the extraordinary history of cyanobacteria. Remarkable metabolic innovators that made breathable air possible, these hardy microorganisms epitomize bacterial continuity in a world of constant change. But the lessons of Spitsbergen seep back into our thoughts, prompting one last question. If Bil'yakh cherts are full of cyanobacteria, what might we find in associated shales formed on the open seafloor? In Spitsbergen, comparable rocks brim with the fossils of eukaryotic algae and protozoans. Do they occur in Bil'yakh shales, as well?

Alexei Veis, a colleague of Semikhatov and Sergeev in Moscow, has studied these shales thoroughly, finding cyanobacterial filaments along with hollow organic balls compressed into minute platters. We can't be certain, but the large size of these balls (up to 500 microns) suggests that they are indeed the remains of eukaryotic cells. Rocks of similar age in Australia contain rare microfossils that are undoubtedly eukaryotic, their biological affinities given away by long, branching arms that ornament cell walls. Nonetheless, rocks of this age do not contain the exuberantly spiny spore coats, vase-shaped microfossils, or many-celled seaweeds found in Spitsberegn shales. By 1.5 billion years ago, then, the cyanobacterial revolution may have been complete, but a second revolution—the rise of eukaryotes to ecological prominence—was yet to come.

8 | The Origins of Eukaryotic Cells

Bacteria may have evolved by exchanging genes, but eukaryotes went them one better. Chloroplasts and mitochondria, the seats of energy metabolism in eukaryotic cells, arose by the lateral transfer of entire cells. Electron microscopy and molecular biology illuminate many aspects of eukaryotic cell evolution, but we still don't understand how our own domain originated.

Beyond our distantmost candle of fact lies a seductive darkness. Scientists are drawn to this blackness because we know it hides more candles, as yet unlit. We strike matches of hypothesis in the hope that a new wick will catch fire. Hypotheses seek to explain what we know, but more important, they make predictions about what we don't know—about experiments not yet run, or fossils not yet discovered. For this reason, hypotheses provide built-in criteria for evaluation: do they help us to light the next candle or not?

Most hypotheses turn out to be wrong—some gloriously and others ignominiously so. This isn't because scientists are dim or the exercise futile. It simply reflects the difficulty of fashioning a lasting explanation of nature. In fact, most hypotheses include useful ideas that survive to become part of the next model or scenario. Good hypotheses also spur new research and so provide value even when the research shows them to be flawed. Most of us develop hypotheses destined for modest success or failure, but on rare occasions an idea comes along that changes how we think about nature. Konstantin Sergeevich Merezhkovsky made one such proposal.

Merezhkovsky, professor of botany at Kazan University, hypothesized in 1905 that the cells of algae and plants are chimeras made up of two originally independent organisms united in obligatory and permanent partnership. Specifically, Merezhkovsky proposed that the chloroplast—the seat of photosynthesis in eukaryotic cells—originated as a cyanobacterium that was swallowed by a protozoan. In framing this hypothesis, he was attempting to explain an observation made years earlier by the German botanist A.F.W. Schimper. Pushing the limits of nineteenth-century microscopy, Schimper had observed that chloroplasts grow and divide independently of (although in synchrony with) the cells that surround them. Just as Pasteur had established that all life springs from life, Schimper showed that chloroplasts lost from cells cannot be generated anew—chloroplasts spring always and only from chloroplasts. Merezhkovsky was also familiar with research showing that corals and some other animals harbor symbiotic algae in their tissues. Insightfully, he combined these two observations to reach his remarkable conclusion: "Chlorophyll bodies grow, are nourished, synthesize proteins and carbohydrates, hand down their characteristics—all independent of the nucleus. In a word, they behave like independent organisms and should be examined as such. They are symbionts, not organs."

Merezhkovsky's hypothesis of endosymbiosis (two cells yoked in mutually beneficial partnership, with one cell nested inside the other) excited vigorous debate in its time, but eventually faded from view, undercut by neglect as much as by experimental refutation. Questions accumulated faster than answers—how did the symbiont become established inside the host's cytoplasm? how did the symbiont fall under the genetic spell of the nucleus?—and in the absence of compelling explanations, biologists moved on to more tractable problems. By the early 1960s endosymbiotic theory drew mention in an American textbook only as "a bad penny that has been in circulation far too long." In Soviet encyclopedias, Merezhkovsky was remembered for his contributions to systematic botany; elsewhere he was rarely remembered at all.

In the fall of 1972, I was an undergraduate looking for a term paper topic in botany. Sensing my need for direction, my professor steered me toward some then recent publications by Lynn Margulis, a young cell biologist with radical ideas. In a 1967 paper that I later learned was rejected fifteen times before finding a home in the *Journal of Theoretical Biology*, she (as Lynn Sagan) reinvented the endosymbiotic hypothesis for eukaryotic cell origins. (Lynn wasn't consciously resuscitating the theories of Merezhkovsky; in 1967 she had never heard of him.) Lynn proposed not only that chloroplasts had originated as endosymbiotic cyanobacteria, but also that mitochondria, the compartmentalized sites of respiration in eukaryotic cells, were descended from free-living, respiring bacteria.

In Darwin's great vision, evolution is fundamentally a process of branching, of divergence—new forms and physiologies arise as the descendants of a common ancestor grow ever more different from one another. Lynn Margulis, however, argued for the emergence of evolutionary novelty as branches *fused*. In her view, each cell in my body reflects the union of two genetic lineages; the rose bush outside my window, blessed with chloroplasts as well as mitochondria, combines three distinct lines of descent. For me, the electric effect of Lynn's paper was both immediate (I found my term paper topic) and lasting (early life, I decided, was the field for me).

Today, biologists accept the endosymbiotic origins of chloroplasts and mitochondria as fact, and Lynn Margulis owns the National Medal of Science. But why did she succeed where Merezhkovsky had failed? Simply put, biologists of the late twentieth century had tools at their disposal that earlier generations could not have imagined. With electron microscopy came the observation that chloroplasts and cyanobacteria share a common structural organization. Biochemistry further showed that cyanobacteria and chloroplasts are nearly identical in the molecular details of photosynthesis. Moreover, chloroplasts were found to respond to antibiotics as bacteria do, not like the nucleus and cytoplasm of the cells in which they occur. Perhaps most surprisingly, chloroplasts turned out to contain DNA, RNA, and ribosomes—the basic molecular machinery for cellular growth and replication.

In tandem, electron microscopic and biochemical research uncovered another striking feature of algal cells. Chloroplasts in red and green algae (and their descendants, the land plants) are surrounded by two membranes. The outer membrane is synthesized by the surrounding cytoplasm, following genetic instructions issued from the nucleus. In contrast, the inner membrane is made by the chloroplast itself. Moreover, the outer chloroplast membrane is part of an extensive membrane system that includes the cell's bounding membrane, the nuclear membrane, and an internal membrane system that permeates the cytoplasm. These membranes are in dynamic continuity, which means that while they may be distinct and unconnected at any one moment, they occasionally combine to form a complex and nearly continuous surface. The significance of this seemingly arcane detail is that the nucleus and cytoplasm lie within this membranous boundary, whereas the chloroplast and its inner membrane lie *outside* it (figure 8.1).

Lynn Margulis argued that these observations can all be explained by the endosymbiotic hypothesis, forcing biologists to reevaluate an idea long dismissed. The clinching evidence, however, came from molecular biology. As we saw in chapter 2, comparisons of nucleotide sequences in genes provide a powerful means of determining evolutionary relationships among organisms. Knowing this, we can construct an elegant test

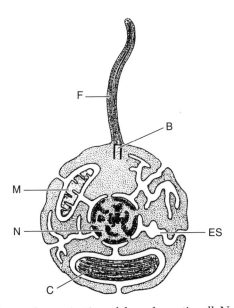

Figure 8.1. The internal organization of the eukaryotic cell. Note that the membranes of eukaryotes, including the endomembrane system (ES), define a space that contains the nucleus (N) and cytoplasm. Chloroplasts (C) and mitochondria (M), however, lie outside this space. Diagram also shows uniquely eukaryotic flagellum (F) anchored by a basal body (B). (Adapted from a figure by Max Taylor)

of the endosymbiotic hypothesis. Merezhkosvky's and Margulis's proposal is fundamentally genealogical—an evolutionary merger uniting microbes from two distinct domains of life. If the hypothesis is correct, gene sequences from chloroplast DNA should be more similar to those of cyanobacteria than they are to genes in the nuclei of plant and algal cells. This turns out to be the case; in the Tree of Life, chloroplasts nestle among the cyanobacteria.

Merezhkovsky was right. Half a century later, so was Margulis. And the humble cyanobacteria take on a new importance as the source of photosynthesis in plants and algae. When you admire the greenery of a tropical forest, you are seeing blue-greens propelled to unprecedented ecological success by hitchhiking in a protozoan.

How can one cell become an integral part of another? The first requirement is straightforward: the host must not digest its guest. A cyanobacterial symbiont must have generated some product that inhibited release of its host's digestive enzymes. The substance was sugar, leaked from the endosymbiont and absorbed by the cell that surrounded it. Successful host cells facilitated photosynthetic sugar production by ensuring a steady supply of carbon dioxide and nutrients to the symbiont. Through this metabolic exchange, a partnership emerged.

Alliances of this type are actually common in nature. For example, as Merezhkovsky recognized a century ago, reef corals harbor unicellular algae within their tissues, exchanging food for nutrients. The algae make it possible for corals to grow rapidly, but when the coral can't hold up its end of the bargain, the algae depart, leaving their animal host pale—and doomed. In the present-day Caribbean Sea, coral "bleaching" linked to rising temperatures poses a serious threat to reef ecosystems.

Chloroplasts are manifestly *not* free to forsake their hosts. They have become anchored in place by a second type of exchange—one that is genetic rather than metabolic. Chloroplasts contain less than 10 percent of the DNA found in free-living cyanobacteria—in the transition from cell to organelle, the endosymbionts lost most of their genes.

How do chloroplasts function in the face of such genetic impoverishment? The answer is that proteins encoded by nuclear genes and synthesized in the cytoplasm are imported into the chloroplast. This requires proteins called chaperones that ferry molecules across the chloroplast membranes. Chaperone proteins are ancient components of the cell's machinery, originally formed to help new proteins fold properly and later co-opted as a transport service. Given this molecular support system, some cyanobacterial genes became redundant and were lost. And, in a process not well understood, some chloroplast genes actually migrated into the nucleus. In consequence, the photosynthetic factory came under nuclear control. From two distinct genealogical lineages, a new type of organism emerged.

Algae don't all cluster together on the Tree of Life; instead, they scatter across several branches of the eukaryotic limb (figure 8.2). In principle, this spread can be explained in two different ways. Possibly, photosynthesis arose once, early in the evolutionary history of the Eucarya, and was later lost in some lineages, including our own. The alternative is that photosynthesis came to eukaryotes several times, by *repeated* symbiotic events. These two hypotheses make predictions that can be tested

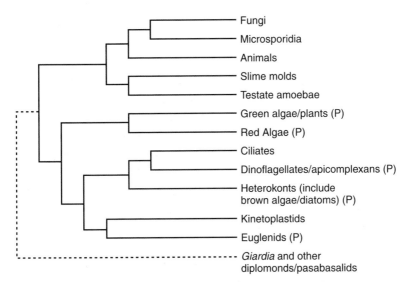

Figure 8.2. A current hypothesis of genealogical relationships among eukaryotic organisms, based on molecular sequence comparisons of ten genes. Note the dotted line that connects diplomonads (which include *Giardia lamblia*) and parabasalids to the remainder of tree. This indicates the uncertainty surrounding the nature and composition of early branches on the tree. Groups with photosynthetic members marked by P. (Redrawn from a figure by Sandra Baldauf)

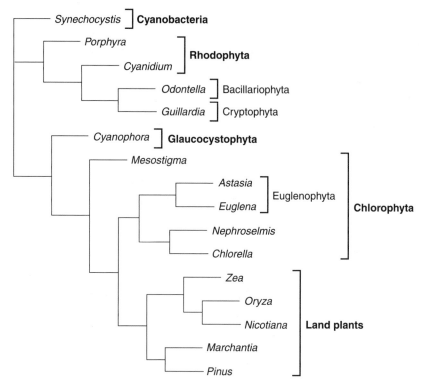

Figure 8.3. Genealogical relationships among chloroplasts, based on molecular sequence comparisons. Note that the chloroplast tree does not show the same relationships as trees based on nuclear gene sequences. This strongly supports the idea that many eukaryotes acquired photosynthesis by engulfing other eukaryotic cells. Photosynthetic euglenids, for example, appear to be derived from endosymbiotic green algae, while the chloroplasts in cryptophyte and heterokont algae (here represented by the Bacillariophyta, or diatoms) seem to be descended from red algal symbionts. (Phylogenetic data courtesy of Paul Falkowski)

by molecular sequence comparisons. If all algae descended from a single symbiosis, then evolutionary trees based on comparisons of chloroplast genes should show the same genealogical relationships as trees based on nuclear genes. They don't. Comparison of figure 8.3, which shows a molecular phylogeny of chloroplast genes, with the tree of eukaryotic organisms depicted in figure 8.2 tells us that endosymbiosis must have brought photosynthesis to eukaryotes half a dozen times. What's more, the story has a twist.

The cryptophytes are a small group of single-celled algae found in temperate and high-latitude waters. In a groundbreaking electron microscopic study, Sarah Gibbs of McGill University showed that cryptophyte chloroplasts are surrounded by four membranes, not just the two found in red and green algae. Moreover, a small dark body called the nucleomorph lies between the inner and outer membrane pairs. Surprisingly, Gibbs found that the nucleomorph contains DNA. These observations suggested to her that cryptophytes, like red and green algae, gained photosynthesis endosymbiotically. But in the case of cryptophytes, she hypothesized, the symbiont was itself a eukaryotic alga, not a cyanobacterium. Gibbs reasoned as follows: the two innermost membranes surrounding cryptophyte chloroplasts represent the two chloroplast membranes of the engulfed algal symbiont; the nucleomorph and associated material between membrane pairs are remnants of the symbiont's nucleus and cytoplasm; and the two outer membranes reflect the symbiont's cell membrane and the enveloping membrane synthesized by its captor. Gene sequence comparisons confirm Gibbs's hypothesis: the chloroplast genes of cryptophyte algae cluster with those of cyanobacteria, genes retained in the nucleomorph branch with nuclear genes of red algae, and genes of the cryptophyte nucleus represent a third distinct eukaryotic lineage. Cryptophytes did indeed acquire photosynthesis by swallowing a eukaryotic alga.

Figure 8.4 summarizes the spread of photosynthesis through the Eucarya. The chloroplasts of red algae have photosynthetic pigments similar to those of cyanobacteria and, as already noted, are surrounded by two membranes; they reflect the *primary* endosymbiotic incorporation of a cyanobacterium by a eukaryotic cell. Green algal chloroplasts have somewhat different pigments—chlorophyll b was added and proteinaceous pigments lost—but their double membrane also reflects primary endosymbiosis. Given increasing evidence that red and green algae are closely related, a single symbiotic event may account for both groups.

Only two groups of algae are known to contain nucleomorphs. Nonetheless, biologists agree that algae other than reds and greens acquired photosynthesis via *secondary* endosymbioses that incorporated eukaryotic symbionts, a conclusion supported by sequence comparisons of chloroplast genes and studies of membrane organization. There is even one case of *tertiary* endosymbiosis, found within the dinoflagel-

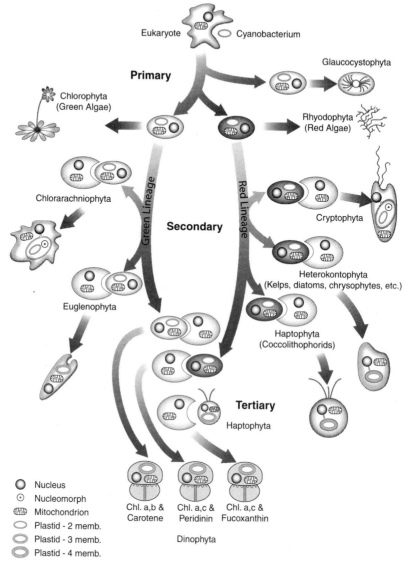

Figure 8.4. A summary of the endosymbiotic events by which photosynthesis spread through the Eucarya. (Reproduced with permission from an illustration by Charles Delwiche)

lates, common members of marine plankton (and the photosynthetic symbionts of corals). These biological Matryoshka dolls originated when a flagellated protist engulfed a so-called haptophyte alga—which itself arose when a protozoan swallowed a unicellular alga closely related to living reds—which, in turn, evolved via the endosymbiotic incorporation of a cyanobacterium into a eukaryotic host! Lewis Thomas once wrote, "I take it as an article of faith that [the committee] is the most fundamental aspect of nature that we know about." Dinoflagellates epitomize this view of life.

How surprised should we be by the spread of photosynthesis through the eukaryotic domain? On first encounter, it seems almost implausibly odd—an improbable sequence of events completed in some distant, more permissive ocean. But, as noted earlier, the symbiotic acquisition of photosynthesis has occurred repeatedly through time. The lichens found on hillside rocks comprise symbiotic associations between fungi and green algae or cyanobacteria. Some of these associations are casual, and the two partners can be grown separately; others have become so intimate that segregation of host and symbiont is no longer possible.

Like reef corals, *Tridacna*, the giant clam of tropical oceans, farms microscopic algae within its tissues. Algal symbionts occur as well in flatworms, sponges, and untold protozoans (including ciliates, radiolarians, and foraminifers). Even some sea squirts, distant relatives of our own, harbor symbiotic cyanobacteria. If you want to see something *truly* unusual, look in the mirror—vertebrate animals don't seem to be capable of forming symbioses with photosynthetic microorganisms.

As Lynn Margulis recognized, the story of mitochondria parallels that of chloroplasts. Just as photosynthesis in eukaryotic cells is localized within chloroplasts, aerobic respiration—the metabolism that fuels our own bodies—is confined to mitochondria. In structure and biochemistry, these small organelles closely resemble members of a bacterial clade called the proteobacteria. Like chloroplasts, mitochondria are bounded by two membranes, and they also contain DNA, RNA, and ribosomes. As early as 1925, Ivan Wallin, a professor of anatomy at the University of Colorado, proposed that mitochondria are fundamentally bacterial. He even claimed that he could culture them as free-living organisms. No one else could do this (in fact, it can't be done), and Wallin, like Merezhkovsky before him, faded into obscurity. But once again, molecular biology showed that Wallin was not wrong (at least not in his general thesis); he was just ahead of his time. Molecular phylogeny shows clearly that mitochondria originated as bacterial cells that evolved through time from casual symbionts to obligatory organelles. In this case, the encompassing host supplied the sugar; the protomitochondrial symbionts returned energy (in the form of ATP) in copious supply.

In chapter 2, we compared the dazzling metabolic diversity of prokaryotic organisms with the limited capabilities of eukaryotes. Nucleated organisms may photosynthesize, respire aerobically, or, occasionally, ferment, but that's about the extent of their abilities. Our modern understanding of mitochondria and chloroplasts makes this comparison seem even more lopsided, as the two metabolisms that power most eukaryotic cells can be traced to lateral transfer on a grand scale—the wholesale importation of bacterial cells. It reminds me of a scene in *A Streetcar Named Desire*, in which Blanche DuBois declares that she has "always benefited from the kindness of strangers." Eukaryotes are the Blanche DuBois of biology.

If chloroplasts and mitochondria originated as bacterial symbionts, how did the rest of the eukaryotic cell take shape? Recall from chapter 2 that eukaryotes share a number of fundamental characters not found in archaeans or bacteria. Their defining feature, of course, is the nucleus: the membrane-bounded compartment that contains the cell's genes. Inside the nucleus, long strands of DNA are wound tightly around tiny proteinaceous beads to form linear chromosomes. The electron microscope reveals an intracellular landscape that further differentiates eukaryotes from other organisms (figure 8.1). Nucleated cells have a distinctive arrangement of proteinaceous strands called microtubules in their flagella, their cell functions are localized within well-defined organelles such as the Golgi apparatus (flattened sacs involved in intracellular transport and secretion), and, as we've already seen, they have mitochondria and chloroplasts. In detail, eukaryotes are also biochemically unique: transcription, translation, and ribosome structure are all distinct from comparable features in prokaryotes.

Perhaps the most important differences between eukaryotes and other cells concern the way in which the cell's contents are stabilized. Archaeans and bacteria enclose their cytoplasm in a rigid wall. In contrast, eukaryotes evolved an internal scaffolding called the cytoskeleton, and that, as Robert Frost once wrote, has made all the difference. Built from tiny filaments of actin and other proteins, the cytoskeleton is a remarkably dynamic structure, continually able to form and re-form in ways that change the cell's shape. Many of us remember film clips of amoebas, viewed on slow days in high school biology class. The graceful undulations of the amoeba, its pseudopodia extending to capture prey, reflect the coordinated action of a dynamic cytoskeleton and a flexible membrane system. This coordination is key to the evolutionary success of eukaryotes precisely because it enabled these cells to engulf particles—paving the way for the acquisition of mitochondria and chloroplasts.

Endosymbiosis may explain the biology of mitochondria and chloroplasts, but it does little to illuminate the many other attributes of eukaryotic cells. Indeed, in terms of classical endosymbiotic theory, the cell that swallowed the protomitochondrion must already have been a card-carrying eukaryote with respect to other key features. This view of eukaryotic evolution makes an explicit genealogical prediction. Eukaryotes that contain chloroplasts and mitochondria should be distal branches on a limb that also includes cells with mitochondria alone. (No chloroplast-bearing cells lack mitochondria.) And the earliest branches of the eukaryotic tree should contain nucleated cells without mitochondria—the prototypical eukaryotes that set all the endosymbioses in motion.

During the 1980s, Mitchell Sogin of the Marine Biological Laboratories in Woods Hole, Massachusetts, pioneered molecular studies of eukaryotic phylogeny. Inspired by Carl Woese, Sogin turned to sequence comparisons of ribosomal RNA genes to sort out eukaryotic relationships. Remarkably, the trees he constructed fit the genealogical predictions of the endosymbiotic hypothesis beautifully: algae reside on upper branches, middle branches include amoebas and other cells with mitochondria but no chloroplasts,[1] and most intriguing, the lowermost

---

[1] Euglenids, a group that includes tiny green flagellates, are an exception, but molecular biology and ultrastructure tell us that euglenids acquired photosynthesis by the endosymbiotic incorporation of green algal chloroplasts. Thus, their ability to harness sunlight must postdate the emergence of green algae.

branches contain nucleated cells with neither mitochondria nor chloroplasts. *Giardia lamblia*, an intestinal parasite well known to backpackers who drink untreated water, serves to illustrate these early branching eukaryotes. *Giardia* is a tiny bullet-shaped cell that thrives in the oxygen-poor environments provided by vertebrate digestive tracts. It has a relatively simple internal organization—for example, there is only limited evidence of a Golgi apparatus—and it retains details of DNA transcription otherwise known only from prokaryotic organisms. In many ways, these cells provide an excellent model for an early eukaryotic cell. But there is a problem. While it is attractive to interpret *Giardia*'s simple biology in terms of ancient origins, we might instead view its peculiar features as adaptations for a parasitic mode of life. Because they obtain many of their physiological needs from their hosts, parasites commonly display a stripped-down biology. Perhaps, then, *Giardia* once had mitochondria but lost them. As it turns out, *Giardia* has some free-living relatives that share many of its unusual traits, so parasitism doesn't explain everything. Nonetheless, because *Giardia*'s free-living cousins also live in oxygen-depleted habitats, the question persists: are eukaryotes that lack mitochondria primitively simple, or were their mitochondria jettisoned as unnecessary baggage in the anoxic environments where these cells thrive?

This question might seem intractable, but in fact there is an elegant means of addressing it. Recall that as endosymbionts evolved into organelles, some of their genes were transferred to the nuclear genome. This being the case, we can ask about lost mitochondria in a relatively straightforward way: do the nuclei of mitochondria-free eukaryotes contain genes transferred from since-departed endosymbionts?

The answer, in some cases at least, is yes. Genetic vestiges of lost mitochondria were first discovered in *Entamoeba histolytica*, an anaerobic parasite nested among conventional amoebas in the eukaryotic tree. Specifically, the nuclear genome of *E. histolytica* contains a gene for a chaperone protein called cpn60 that is closely related to the cpn60 genes of mitochondria and free-living proteobacteria. The simplest explanation for this observation is that entamoebas had mitochondria and lost them—but they didn't lose all their mitochondrial *genes*.

In the wake of this discovery, a number of molecular biologists, including Mitch Sogin, turned their attention to *Giardia* and other lineages that branch early in trees based on ribosomal RNA genes. *All* contain nuclear genes of proteobacterial origin. Thus, all known eukaryotic cells show evidence of early symbiosis with bacterial cells. Either the endosymbiotic incorporation of mitochondria preceded the last common ancestor of all living eukaryotes, or early eukaryotes hosted a number of since-departed proteobacteria as symbionts or parasites. The latter possibility isn't so far-fetched. We know from microbiological experience that proteobacteria—which include *E. coli*, the bacteria that fix nitrogen in soybean roots, and the chemosynthetic microbes that nourish tube worms along deep-sea rift vents, not to mention the bacterial vectors for Legionnaires' disease, typhus, and Rocky Mountain spotted fever—are particularly good at taking advantage of the biological habitats offered by eukaryotes. Inserting genes for chaperone proteins into the host's nucleus may be a widespread way of dropping proteobacterial anchor in a new biological harbor.

Regardless of how we interpret these migratory genes, comparative biology appears to leave us with few clues to the origins of eukaryotic cell biology. Or perhaps the clues are hiding in plain sight, awaiting a new Lynn Margulis to make sense of them. In a provocative essay published in 1998, William Martin and Miklos Müller hypothesized that primitively mitochondria-free eukaryotes never existed. Instead, they proposed, eukaryotic cell organization originated in a primordial symbiosis between two prokaryotes. One partner was a methanogenic archaean needing H_2 and CO_2 for fuel. The other was a proteobacterium able to respire aerobically in the presence of oxygen, but also able to live anaerobically by means of fermentation—producing H_2, CO_2, and acetate as waste products. Blessed with complementary metabolisms, the two partners united to form a microcosmic carbon cycle. Organic molecules produced by the methanogen were imported into the proteobacterium; the proteobacterium, in turn, provided the methanogen with the hydrogen and carbon dioxide it needed to produce more organic matter. As hydrogen levels declined in the ocean and atmosphere, driven in part by the great Proterozoic oxygen revolution, the methanogens clung ever closer to their partners, eventually jettisoning their walls and evolving flexible membranes that enabled them to maximize hydrogen gain by surrounding their bacterial confederates. In the absence of walls, new, internal means of stabilizing cell contents were needed—and were accomplished by the evolution of cytoskeleton proteins. Genes were transferred or lost, and a new cellular organization emerged.

Martin and Müller were not the first to suggest that nucleated cells originated by primordial symbiosis, but their hypothesis is distinguished by its ecological logic as well as its phylogenetic predictions. It explains why all known eukaryotic cells contain proteobacterial genes. Further, it provides a rationale for the observation that bacterially derived genes in eukaryotes tend to be related to metabolism, while genes most closely related to those of archaeans commonly function in transcription and translation. (Note that in this view, the position of Eucarya in the Tree of Life tells us only about the ancestry of eukaryotic RNA genes; the origin of eukaryotes as new evolutionary entities requires a fusion of lineages that can't be captured in full by phylogenies based on single genes.)

The hypothesis may even help to explain why proteobacteria have been so successful in establishing pathogenic and mutually beneficial relationships with eukaryotic organisms. Perhaps these organisms can exploit eukaryotic hosts because they recognize distantly shared genes.

The Martin-Müller hypothesis addresses one other, seemingly esoteric feature of eukaryotic cell biology. As noted earlier, eukaryotes that live in oxygen-free environments lack mitochondria. But some of them harbor another organelle, called the hydrogenosome, that directs anaerobic metabolism in these cells. More than twenty years ago Miklos Müller proposed that hydrogenosomes are, like mitochondria, energetic organelles derived from bacterial symbioses. That was a bold assertion because hydrogenosomes do not contain DNA. Thus, if they are descended from free-living bacteria, hydrogenosomes must have surrendered *all* of their genes. This claim seems preposterous, but, remarkably, increasing evidence suggests that it is correct. Studies of *Trichomonas*, a mitochondria-free parasite that contains hydrogenosomes, show that its nuclear genome contains several genes of proteobacterial origin. Moreover, the proteins encoded by these genes do their work in the hydrogenosome. Not only do hydrogenosomes appear to be proteobacterial symbionts reduced to gene-free metabolic slaves, but sequence comparisons of their telltale genes suggest that these organelles are most closely related to—mitochondria!

Despite these virtues, the Martin-Müller hypothesis can't explain everything we know about eukaryotic cells. In recent years, scientists have begun to determine the nucleotide sequences for entire genomes—that is, for all the DNA in an organism, not just single genes. As more and more complete genomes become known, it is becoming possible to identify those genes that are shared universally and those that occur only in a specific domain.[2] Hyman Hartman and Alexei Fedorov, molecular biologists at MIT and Harvard, respectively, have identified hundreds of eukaryotic genes that do not occur in Bacteria or Archaea and, hence, appear to be molecular signatures of eukaryotic biology. Hartman and Fedorov contend that these genes identify a third partner in the aboriginal symbioses that led to eukaryotic cells, an early life-form that may be represented today only by the genes it contributed to eukaryotic cells. Not surprisingly, many of these genes relate specifically to the signature cellular features of eukaryotes—the cytoskeleton and nucleus.

Clearly, we are a long way from resolving all mysteries of eukaryotic cell origins. But hypotheses like those of Martin and Müller, and of Hartman and Fedorov, cap a strengthening view of early evolution in which nature appears not so much "red in tooth and claw" as "green in mergers and acquisitions"—a perspective as appropriate to twenty-first-century economics as competition and survival were to Victorian capitalism. These hypotheses are stimulating, radical, and provocative, just as Lynn Margulis's thesis was in 1967. And, like Margulis's ideas, they are catalyzing new research that promises fresh insights into one of biology's deepest riddles. That's what a good hypothesis does.

A POSTSCRIPT

Molecular investigations of eukaryotic phylogeny are currently in high gear. New genes are being sequenced, and analytical methods are improving. Equally important, biologists are sampling genes from an increasing large subset of eukaryotic diversity. The final word on eukaryotic

---

[2] As I write this, more than fifty species have been sequenced completely, and many more are in the pipeline.

phylogeny is not yet in, but as shown in figure 8.2, increasing evidence suggests that animals are closely related to fungi (how about *that* for our family tree), and that animals + fungi are further allied with amoebas and slime molds. Another branch, still contentious, unites red and green algae, supporting the thesis that the primary endosymbiosis between cyanobacteria and protozoans occurred only once. Other groupings unite apparently strange bedfellows: heterokont algae (which include kelps and diatoms) share a branch with the funguslike oomycetes, while ciliates, dinoflagellates, and plasmodia (the infectious agents of malaria) group together on a nearby limb. Perhaps most riveting, some bugs that occupied basal branches in Mitch Sogin's early RNA-based trees have been relocated to higher limbs as more genes weigh in. Notably, the microsporidia, tiny parasites that branch near the base of the RNA tree, nest with the fungi in genetically more inclusive phylogenies. Evidently, rapid rates of evolution caused the ribosomal RNA genes of microsporidians to be quite different from those of other eukaryotes, forcing them toward the bottom of trees based on RNA gene similarity. The same may be true of some other parasites, but *Giardia* and the hydrogenosome-bearing trichomonads may persist as early branching protists. The tree is still growing and taking shape. But we now know enough that we can return to the fossil record to ask how eukaryotic evolution is reflected in Proterozoic rocks.

9 | Fossils of Early Eukaryotes

Eukaryotic organisms have evolved patterns of cell shape and multicellularity unknown in bacteria and archaeans. Fossils with these features suggest that eukaryotes arose early, but emerged as prominent participants in marine ecosystems only late in the Proterozoic Eon, perhaps aided by renewed oxygen increase in the world's oceans.

THE YOUNG MINER taps insistently on my shoulder, gesturing vigorously and shouting (in Chinese) what must surely be instructions. He points toward a nearby truck—half a dozen workers have already dived beneath it. Being no fool, I follow their example, and seconds later an explosion jolts the ground, followed by a hail of rock debris that dents the cab above my head.

We are in Guizhou Province in southern China, visiting one of the many phosphate mines that pock the rugged landscape (figure 9.1). Phosphates find use in fertilizer, and the Guizhou mines play an integral role in the regional economy. What the miners don't know is that Guizhou phosphate, spread on fields from Kunming to Kalamazoo, contains some of the most exquisite fossils ever found in Proterozoic rocks. Moreover, the fossils are predominantly eukaryotic. Sometime between the deposition of the Great Wall and these younger phosphatic rocks, nucleated organisms broke the 2-billion-year ecological hegemony of the bacteria. Prokaryotes didn't go away, of course. They remain the foundation of all functioning ecosystems on this planet. But algae joined and then supplanted cyanobacteria as the principal primary producers in the oceans, and protozoans able to engulf microscopic victims added the complexity of predation and herbivory to food webs.

Figure 9.1. Fossiliferous phosphate rocks of the Doushantuo Formation, exposed in a quarry at Weng'an, China.

Guizhou fossils provide a great introduction to the Proterozoic history of eukaryotes.

The Guizhou phosphates lie along the thin edge of a massive wedge of sedimentary rocks deposited in southern China near the end of the Proterozoic Eon. To the north, where they are exposed in the spectacular Yangtze Gorges, these rocks can be divided into four units that lie one atop another. At the base is a discontinuous blanket of red sandstones formed by meandering streams as they traversed a coastal plain; volcanic ash found in a thin layer within the sands yields a U-Pb age of 748 ± 12 million years. At the top of the succession is a thick cover of limestones and dolomites that contains Early Cambrian fossils in its uppermost part. In between are two units of particular interest: one provides a record of extreme climate, while the other contains spectacular fossils that have reshaped our understanding of Proterozoic life.

The lower unit, called the Nantuo Tillite, lies directly above the beds of red sandstone. A poorly sorted mixture of boulders, sand, and silt, this formation is widely distributed in southern China. Running water tends to separate sediment particles of differing size and density, so the intimate mixing of silt and football-size boulders suggests a different means of transport—ice. Other sedimentary features confirm a glacial origin for Nantuo rocks. For example, dropstones—isolated pebbles and cobbles plunged into finely laminated silts and muds—record icebergs that rafted coarse debris out onto the ocean before melting and dropping their rocky cargo onto fine-grained sediments below. Pebbles in the tillite display deeply incised striations formed by grinding as rock-studded ice moved across the landscape. Glacial rocks can be seen in younger Proterozoic successions around the world. As we shall see in chapter 12, they provide evidence for a series of ice ages so severe that life itself may have hung in the balance.

As the glaciers melted, rising seas began to deposit the second unit of interest—the fossiliferous Doushantuo Formation. In the Yangtze Gorges region, nearly 1,000 feet of shale, phosphatic rocks, and carbonates accumulated during two cycles of sea-level rise and fall. To the southwest, closer to the ancient shoreline, the formation thins and changes character. In Guizhou, where I got my lesson in mining safety, it is only 140–160 feet thick and consists mainly of phosphatic rocks deposited in near-shore marine environments. Only a few miles farther west, a mere 16 feet of phosphatic sandstone document this interval. To date, no volcanic rocks have been discovered in these beds, but experimental dating based on radioactive uranium and lutetium locked into phosphate crystals as they formed suggests an age of 590 to 600 million years. Encouragingly, this age falls within the range of estimates for Doushantuo deposition (younger than 600 million years and older than

555 million) based on the correlation of its fossils and chemical signatures with those of well-dated successions elsewhere.

The eukaryote-rich rocks of the Doushantuo Formation, then, are much younger than the Great Wall dolomites, and they postdate the fossiliferous rocks of Spitsbergen, as well. Indeed, these rocks are only 50 to 60 million years older than the Cambrian cliffs along the Kotuikan River.[1] Perhaps, then, Doushantuo fossils will not only illuminate the rise of nucleated organisms but hint, as well, of further biological transformation about to begin.

How do we identify a fossil as eukaryotic? For plants and animals, the distinction is easy—no bacteria or archaeans build anything like a leaf or a shell. Microscopic fossils, however, can be more challenging. Biologists find it easy to distinguish between prokaryotes and eukaryotes based on myriad features of cellular organization, genetics, and physiology, but none of these features is available to paleontologists. We have to rely on form.

When Precambrian research was young, paleontologists tried hopefully to recognize eukaryotic microfossils on the basis of size or preserved features of cell biology. Neither worked. It is easy to see the attraction of size—on average, eukaryotic cells are larger than bacteria, and diameter is simple to measure. At the extremes of the scale, size can indeed be informative—bacterial cells more than a millimeter long are unknown, and neither do we know of eukaryotes only 300 nanometers across.[2] But at intermediate sizes (commonly encountered in Proterozoic rocks) there is strong overlap between bacteria and eukaryotes. Tiny green algae in the open ocean are less than a micron in diameter, and—

---

[1] My friend Dick Bambach objects to the word "only" in this sentence, reminding me that 50 million years is a very long time. Indeed it is. In 50 million years we could run the history of Egypt, from the pyramids to modern Cairo, more than 10,000 times. Two million human generations could come and go, and so could more than a billion generations of amoebas. My sentence is meant to convey the fact that *relative* to the enormous span of time that separates Great Wall and Doushantuo deposits, 50 million years is pretty short. But Dick's point is a good one. Every now and again, we should sit back and contemplate the immensity of the canvas on which life's early history is painted.

[2] A nanometer is 10^{-9} meter, or one-thousandth of a micron. A 300-nanometer cell would thus be less than a third of a micron long.

recalling those cigar-shaped fossils in Kotuikan cherts—the resting cells of cyanobacteria can be well over 100 microns long.[3] Cyanobacteria that form extracellular envelopes complicate the picture further, because a colony of 10-micron cells can be enclosed in a preservable coat 100 microns across.

If size isn't foolproof, what about preserved details of cell biology? One of the first Proterozoic fossil assemblages to be discovered was that of the 830–810-million-year-old Bitter Springs Formation in central Australia. Bitter Springs fossils occur in chert nodules within carbonates deposited in ephemeral lakes on an arid coastal plain. The cherts contain beautiful cyanobacteria described by UCLA's Bill Schopf, as well as simple spherical fossils about 10 microns in diameter. Some of these spheres are hollow and were originally interpreted as cyanobacteria. Others, although essentially identical, contain small, dark inclusions of organic matter and so were interpreted as eukaryotic algae with preserved nuclei. Nuclei are mostly water, along with highly nutritious proteins and nucleic acids. In consequence, they are quickly and completely obliterated soon after death—so completely that in the entire fossil record only a handful of plausible fossil nuclei have ever been identified. On the other hand, decomposing cyanobacteria and algae commonly contain small balls of organic matter formed as cell contents shrivel. Decaying cytoplasm provides a satisfactory explanation for the "black spots" in Bitter Springs and other Proterozoic microfossils. Some of these fossils may be eukaryotic, but (unlike leopards) we can't tell them by their spots.

What really makes eukaryotic fossils stand out is *morphology*. In chapter 3, we noted that some cyanobacteria have cell shapes and colony forms not duplicated by other bacteria. Similarly, some (but not all) eukaryotic cells display features that are unknown in prokaryotic organisms. Doushantuo fossils illustrate how paleontologists recognize and interpret early fossil eukaryotes.

In and around the Yangtze Gorges, chert nodules occur at two levels within the Doushantuo Formation. The lower cherts are richly fossil-

---

[3] The largest bacteria currently known are sulfur-oxidizing cells found in sediments off the coast of South Africa. These giants reach diameters of 500 microns or more, although, in a way, they cheat—the cells are hollow.

iferous, mostly preserving tightly interwoven populations of mat-building cyanobacteria. The upper cherts also contain cyanobacteria, but additionally include distinctively different microfossils. These latter remains are broadly spherical, they are commonly large (up to 600 microns), and—most telling—they display flamboyant ornamentation. Some look like tiny suns, with raylike arms extending in all directions (figure 9.2). Others are festooned with spines, flanges, or knobs. Bacteria don't produce structures like this, but several groups of eukaryotic organisms do. Thus, we have confidence that, using Doushantuo fossils as handholds, we can climb our own branch of the Tree of Life.

The exact biological relationships of these exquisite fossils are unknown, but most appear to be the discarded spore coats of algae. Similar remains are known from Australia, Siberia, Scandinavia, and India; they record a global diversification of marine life in the aftermath of widespread glaciation. Remarkably, this diversification was short-lived. For reasons still under debate (but possibly tied to one last expansion of Proterozoic ice sheets), nearly all of these exuberantly ornamented fossils disappeared within a few million years of their appearance—victims of one of Earth's earliest-known mass extinctions.

There is more treasure in Doushantuo rocks. In 1990, the Chinese paleontologist Chen Menge discovered a second, very different fossil assemblage in black shales high up in the cliffs that line the Yangtze Gorges—fossils large enough to be seen by the naked eye, preserved in great numbers as organic films compressed onto bedding surfaces (plate 5a). Once again, distinctive morphologies indicate that many of the thirty or so populations found so far in these rocks represent eukaryotic algae. In some cases, their biological details are so well preserved that we can almost reanimate them. For example, my former student Shuhai Xiao, now at Tulane University, was able to reconstruct common compressions called *Miaohephyton* (literally, the alga from Miaohe) as seaweeds that formed lawns on the ancient seafloor. Thin, grasslike blades stood erect in the water, anchored by rootlike holdfasts to the muddy bottom—in my mind's eye, I can see them sway gently in slack currents. As these algae grew, they branched now and again by splitting in two at their tips. Reproductive cells formed in wartlike structures that line the upper parts of mature individuals, and disper-

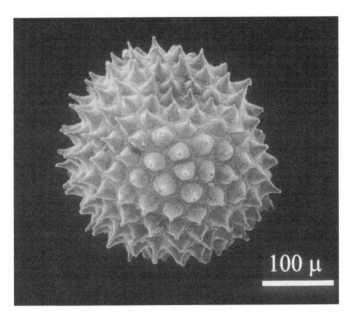

Figure 9.2. A spiny microfossil found commonly in Doushantuo cherts and phosphatic rocks. Such fossils are thought to be the reproductive spores of eukaryotic organisms. Fossil is 250 microns in diameter. (Image courtesy of Shuhai Xiao)

sal occurred both by the release of spores into the water and by the fragmentation of branches along preformed abscission surfaces. A comparable combination of characters can be found today in some brown algae, providing functional and, possibly, genealogical clues to the Doushantuo fossils.

The Maiohe fossils were preserved by rapid burial in fine-grained sediment. Bacteria normally decompose algal tissues soon after death, but at Maiohe they were prevented from completing this task by a veneer of clay that both excluded oxygen and adsorbed the enzymes that break cells apart. As sedimentary layers accumulated, biological remains were pressed like flowers between the pages of a book.

Leaves are commonly compressed in younger mudstones formed in lakes or the floodplains of rivers. In marine rocks, however, organic compressions are rare because burrowing animals irrigate and churn mud and silt layers. Rare, but not unknown. In fact, the most famous of all fossil deposits, the Middle Cambrian Burgess Shale, formed by the compression of animal carcasses in carbon-rich mudstones.[4] Burgess fossils postdate Doushantuo deposition by only 50–85 million years; thus, the Doushantuo compressions are noteworthy for what they lack as well as what they include. One population of flanged tubes may record simple, sea anemone–like invertebrates (plate 5b), but nowhere do we find evidence of the anatomically and morphologically complex animals that are so conspicuous in Burgess and other Cambrian glimpses through the same preservational window. Evidently, much happened between Doushantuo and Burgess time.

I became involved in the study of Doushantuo fossils at the invitation of Professor Zhang Yun, a kind, cultured, and wonderfully insightful paleontologist at Beijing University. During the 1980s, Zhang collected Doushantuo samples from a phosphate mine near the Guizhou village of Weng'an. Several contained multicellular algae. In 1992, to my everlasting good fortune, he asked me to join him in collaborative research. By coincidence, a second Chinese friend, Yin Leiming of the Nanjing Institute of Geology and Palaeontology, invited me to visit China at about the same time. Yin was working on Doushantuo fossils in cherts from the Yangtze Gorges and Zhang was investigating assemblages from the phosphate mines of Guizhou, so I suggested that the three of us join forces to try to understand the full diversity of Doushantuo eukaryotes. Shuhai Xiao, who had completed his bachelor's degree in Beijing under Zhang's guidance, entered Harvard as a graduate student the same year, completing our team.

Doushantuo cherts and shales open two distinct and unusually clear windows on late Proterozoic life, but those Guizhou phosphates provide a third view, even better than the others. In this corner of the shallow Doushantuo seaway, biological remains that entered surface sediments were coated almost immediately by minute crystals of calcium phosphate minerals, preserving both overall morphology and cellular anatomy in remarkable three-dimensional detail. As a result, the Doushantuo phosphates preserve organisms seldom seen in rocks of any age.

---

[4] Discovered by Charles Doolittle Walcott in 1909, the Burgess Shale is renowned for its compressions of Cambrian animal remains. See chapter 11 for discussion.

Like cherts along the Yangtze River, the Guizhou phosphates are chock-full of highly ornamented eukaryotic microfossils; in some beds these fossils are so abundant that they literally form sandstones made of phosphatized cells. Other fossils are multicellular, and once again details of morphology show that most represent tissue-forming algae, not bacterial colonies. Particularly informative are small, crust-forming structures made of thick-walled cells arranged in rows that fan outward, like water as it gushes from a spring (plate 5d and e). Biologists call this anatomical organization a "cell fountain," and it is especially common in red algae. Distinctive reproductive structures and anatomically distinct inner and outer tissues strengthen the ties between these fossils and a particular group of reds called the corallines. (Doushantuo fossils don't display all of the features that collectively define coralline algae, but they exhibit enough of them to suggest that these small phosphatic fossils record an early way station in red algal evolution.) Other Guizhou fossils seem to lie halfway between the two major branches of living red algae, again suggesting that Doushantuo phosphates captured the diversity of red algae *in statu nascendi*—as it began to unfold.

Taken together, compressed and phosphatized Doushantuo fossils show that by the time large animals appeared in the oceans, multicellularity was already well established among the algae.

It is exciting to discover algae with cells preserved intact, but the crown jewels of Doushantuo are undoubtedly small balls, 400–500 microns in diameter, found in phosphates near the village of Weng'an (figure 9.3). The balls are uniform in size. Some contain a single cell wrapped in a thick furrowed coat, while others contain multiple cells surrounded by a thin membrane—cells in pairs, quartets, octads, and larger powers of two arranged in a geometric pattern determined by precisely oriented cell divisions. Parts of this division series were reported by Chinese paleontolgists in 1995 and interpreted as colonial green algae, but size, geometry, and envelope formation collectively make such an interpretation unlikely.

Shuhai Xiao discovered new and more informative populations that enabled him to recognize them as animals—specifically, animal eggs and embryos in the early stages of growth. Fossil embryos are rare in the geological record, but they do occur in Cambrian rocks. In fact, some beautiful Cambrian embryos described a year earlier by Stefan Bengtson

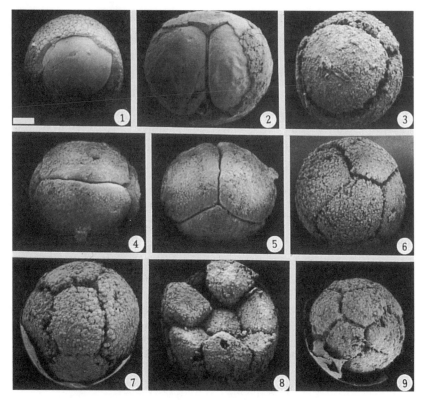

Figure 9.3. Eggs and embryos of early animals, preserved in Doushantuo phosphate. Each fossil is 400–500 microns in diameter. (Images courtesy of Shuhai Xiao)

of the Swedish Museum of Natural History and his Chinese colleague Zhang Yue were Shuhai's inspiration to keep his eyes wide open when sorting Doushantuo fossils.

Later growth stages have not yet been uncovered, so we don't know what kind of adults might have developed from the Doushantuo embryos. Among living animals, arthropods and related invertebrates most closely approximate the egg case and cell cleavage patterns displayed by these fossils, but this doesn't mean that recognizable arthropods plied the Doushantuo seaway. In parallel with Doushantuo shales, Guizhou phosphates contain possible sponges and small tubes likely made by simple coral-like organisms (plate 5c), but they display no evidence of arthropods or any of the other anatomically complex animals found in phosphatized Cambrian rocks. Around 590–600 million years ago, then, animal evolution may have begun, but the age of animals was still to come. Doushantuo fossils preserve the smoldering fuse of an evolutionary explosion about to begin.

The Doushantuo Formation is a paleontological wonder, our closest approximation yet to a Precambrian Burgess Shale. Zhang Yun, the quiet pioneer who gave us so many of these remains, died in 1999, but his students continue to plumb Doushantuo rocks for new fossils and further insights into early evolution. The prospect that this paleobiological mother lode will be exhausted in my lifetime is remote.

Thinking back on the fossils found in Great Wall cherts and shales, it becomes clear that biology changed radically between 1.5 billion years ago, when the Siberian rocks were deposited, and 590–600 million years ago, when Doushantuo sediments formed. And further remarkable events were imminent, even as Doushantuo rocks accumulated in South China. In the following chapters, we will explore what came next. But, for now, the task is to fill the evolutionary gap between the Great Wall and Doushantuo biotas.

Multicellular red and green algae are common in Doushantuo assemblages. As we first learned in chapter 3, green algae closely related to the extant genus *Cladophora* occur in 700–800-million-year-old shales in Spitsbergen. Indeed, microfossils interpreted as spores of planktonic green algae suggest that the "greening" of the oceans began at least a billion years ago.

Red algae also have a long Proterozoic history. The oldest fossils that can be compared with confidence to any living eukaryotes are beautifully preserved filaments found by Nick Butterfield in cherts about 1.2 billion years old from Somerset Island in arctic Canada (plate 6a and b). Each filament in Nick's population consists of aspirin-shaped cells about 50 microns wide, aligned in a row. The cells are defined by thin, dark walls and united by a thicker but lighter outer wall layer. Cells are clearly grouped into pairs and pairs-of-pairs, providing evidence that these organisms grew by cell division within (rather than at the ends of) the filaments. Cells at the basal end are differentiated into holdfasts that anchored filaments to firm sediments on an ancient tidal flat. Another type of cell division is apparent in some filaments, and it is an unusual one; the aspirin-shaped cells sometimes divided repeatedly to form small reproductive bodies that resemble wedges of pie (plate 6b).

Collectively, these features ally the Somerset fossils with simple red algae (plate 6c and d). This means that the reds must have diverged from other eukaryotes, acquired photosynthesis (via endosymbiosis, as explained in the preceding chapter), and evolved a simple form of multicellularity by at least 1.2 billion years ago. Thus, both reds and greens appeared more than a billion years ago and diversified dramatically by 600 to 590 million years ago. Even heterokont algae born of secondary endosymbiosis may have differentiated early on. Fossils from the Lakhanda Formation in southeastern Siberia contain simple branching filaments comparable in morphological detail to the living heterokont *Vaucheria*. Lakhanda beds are cut by (and, so, are older than) igneous rocks dated at 1,003 ± 7 million years.

Other fossils strengthen the view that eukaryotes rose to prominence during the second half of the Proterozoic Eon. For example, microscopic fossils with distinctly eukaryotic spines or other ornamentation first appear in rocks about 1.2–1.3 billion years old and become increasing commonly as we ascend through the late Proterozoic record (plate 6e–g). More specifically, late Proterozoic biomarker molecules and (more controversially) microfossils record the presence of dinoflagellates, members of another major group of eukaryotes. Other biomarkers extracted from 750-million-year-old shales deep within the Grand Canyon suggest the presence of ciliate protozoans—phylogenetic cousins, actually, of the dinoflagellates.

Grand Canyon rocks document the early budding of one more branch on the eukaryotic tree. Distinctive vase-shaped microfossils entered my narrative early, in the discussion of Spitsbergen cherts and shales. Such fossils are common in upper Proterozoic rocks, often occurring in remarkable numbers, and nowhere are they more abundant than in the Grand Canyon. Working with exquisite populations preserved in carbonate nodules just beneath a bed of volcanic ash dated at 742 ± 6 million years, Harvard student Susannah Porter has demonstrated that these tiny vases were constructed by testate amoebas—amoeboid protozoans that live inside a minute shell, or test, of their own making (figure 9.4).

This discovery fascinates me because it sheds light on aspects of late Proterozoic ecology. Most of the microfossils discussed in this and previous chapters document photosynthetic organisms, either cyanobacte-

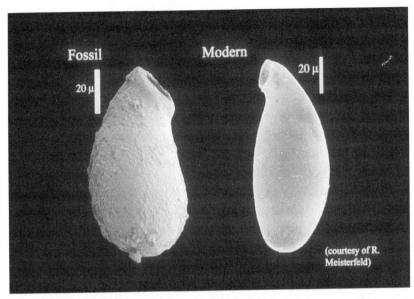

Figure 9.4. Vase-shaped fossil from ca. 750-million-year-old rocks of the Grand Canyon compared with a modern testate amoeba. Note scale on photo. (Figure courtesy of Susannah Porter)

ria or algae. Even the unusual fossils in Gunflint cherts were autotrophic, although they used chemical rather than solar energy to fuel cell growth. In contrast, the vase-shaped organisms were protozoans—heterotrophic eukaryotes that made their living by preying on other microorganisms. The vase-shaped microfossils, thus, tell us of growing ecological complexity in late Proterozoic oceans. Algae and cyanobacteria formed the nutritional base of ecosystems, providing food for untold bacteria. The testate amoebas dined on these algae and bacteria. Moreover, a few vases display hemispherical perforations likely made by other protozoans intent on eating *them*. So, by 750 million years ago, eukaryotes had begun to construct the complex food webs that today form a crown—intricate and unnecessary—atop ecosystems fundamentally maintained by prokaryotic metabolism.

Spiny unicells, multicellular microfossils, compressed macrofossils, eukaryotic biomarker molecules—all can be used to trim the eukaryotic tree with ornaments of time (figure 9.5). They show that as the long Proterozoic Eon moved into its final phase, Earth was becoming a eukaryotic planet.

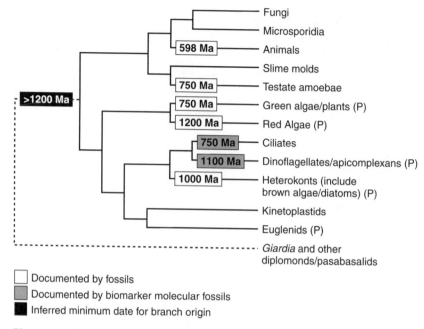

Documented by fossils
Documented by biomarker molecular fossils
Inferred minimum date for branch origin

Figure 9.5. The eukaryotic phylogeny first shown in figure 8.2, here trimmed with the dates of early eukaryotic fossils.

In contrast to the cyanobacteria discussed in previous chapters, most eukaryotic fossils don't range through immense intervals of time. Rather, late Proterozoic algal and protozoan species appear in the record, persist for a discrete period, and then disappear, never to be seen again. The familiar pattern of punctuated equilibrium suggests evolutionary dynamics much like those of younger plants and animals, but enacted at a more leisurely pace—the stratigraphic ranges of many Proterozoic eukaryotes appear to be much greater than those of Phanerozoic species. In general, eukaryotic diversity increased through the late Proterozoic Eon and into the Cambrian, but the progress of diversification was halting, punctuated by several intervals of widespread extinction. As noted at the beginning of this chapter, at least some extinctions appear to be linked to latest Proterozoic climatic change.

There is a practical side to this evolutionary pattern—the eukaryotic microfossils in upper Proterozoic rocks can be used to tell time. Boris Timofeev, another of the Russian geologists charged with sorting out the Proterozoic rocks of Siberia, first recognized this potential. But it was

Gonzalo Vidal, a gregarious Spanish-cum-Swedish paleontologist from Uppsala University, who convinced initially skeptical geologists that Proterozoic eukaryotes came and went in a time-ordered fashion. Beginning with sandstones and shales exposed along the shores of Lake Vättern in south-central Sweden, Gonzalo discovered planktonic microfossils in upper Proterozoic rocks throughout Scandinavia. When he demonstrated how the fossils brought stratigraphic order to these rocks, the modern era of Proterozoic biostratigraphy was born.

As noted in the previous chapter, debate about eukaryotic phylogeny continues, but most disagreement focuses on the identity and character of the tree's early branches. There is widespread agreement that much of the eukaryotic diversity seen today began to accumulate during a relatively short interval of rapid divergence. Paleontology seems to be telling us that this "big bang" of eukaryotic evolution began at least a billion years ago.

If that is true, why did this new type of biology take off so late in the evolutionary day? After all, as noted in chapter 6, sterane molecules extracted from 2.7-billion-year-old shales are thought to be molecular signatures of eukaryotic biology. If the course of eukaryotic evolution was set so early in life's history, why should the domain (*our* domain!) have remained subservient to prokaryotes for a billion and a half years before spreading throughout the oceans? No one really knows, but we can think about four types of explanation. The 2.7-billion-year-old biomarker molecules record only one aspect of eukaryotic biology—the ability to make sterol compounds. Eukaryotes have many other distinguishing features, and perhaps the "complete" eukaryotic cell, with its distinctive genes, differentiated nucleus, cytoskeleton, and mitochondria, evolved much later. Alternatively, eukaryotes could have originated early but diverged much later, in the wake of some enabling environmental event. Or, late divergence might reflect biological innovation—sex is the one most commonly invoked. Of course, we must also ask whether the late Proterozoic radiation might be more apparent than real, reflecting a greater volume of rocks and better fossil preservation rather than increased biological diversity.

Shales that lie beneath the rubbly, tick-infested plains of Australia's Top End discount the first and last explanations. Part of the middle

Proterozoic Roper Group, their age fixed by 1,492 ± 3–million-year-old volcanic rocks, these shales contain microfossils whose abundance and quality match the best preservation seen in upper Proterozoic rocks. Yet there are no spiny fossils like those in Guizhou, no tiny vases like the Grand Canyon and Spitsbergen populations, and no branching compressions comparable to those in younger shales from China or Spitsbergen. In short, Roper assemblages display little of the morphological variety that documents eukaryotic diversity in younger Proterozoic beds. There are, however, eukaryotic fossils.

Most Roper microfossils are large, compressed spheres, much like those in the broadly contemporaneous shales that flank Great Wall carbonates in northern Siberia; they are probably, but not demonstrably, eukaryotic. But one small population discovered by Emmanuelle Javaux, a Belgian postdoc in our lab, provides strong evidence of cytologically sophisticated eukaryotes in mid-Proterozoic oceans. The fossils are moderately large spheroids, about 30–150 microns in diameter, distinguished by one to as many as twenty long, slender tubes that arise from their walls (plate 6e). The tubes are irregular in number as well as position, and they sometimes branch. Similar shapes can be seen in some living protists, where tubes develop as extensions of spore walls, allowing reproductive cells that differentiate inside to escape and disperse. By analogy, then, the irregular tubes on the Roper fossils suggest microorganisms that could modify their shape during the lifetime of a single cell. Bacteria don't do this very well, but eukaryotes are masters of the trade—their ability to form and re-form cell shape is conferred by the cytoskeleton, the dynamic internal scaffolding introduced in chapter 8. This being the case, Roper fossils tell us, not only that eukaryotic microorganisms were present nearly 1.5 billion years ago, but also that they already boasted some version of the sophisticated internal organization seen in living eukaryotes.

In addition to microfossils, Roper shales contain molecular fossils, including steranes that provide complementary evidence for eukaryotic life. Even macroscopic fossils of likely eukaryotic origin occur in middle Proterozoic rocks. Observant paleontologists in Australia, China, India, and the United States have uncovered helical compressions an inch or so across in mid-Proterozoic siltstones as well as short strings of beads 1–3 millimeters in diameter preserved as impressions on the surfaces of sandstone beds. These fossils are hard to classify; conceivably, they record extinct lineages only loosely related to modern eukaryotes.

Collectively, then, paleontological discoveries indicate that the late Proterozoic ascendance of algae and protozoans was not the starting gun for eukaryotic life. Nor is the documented increase in eukaryotic diversity simply an artifact of rock preservation or sampling. *Something*—some biological innovation or environmental shift—must have happened to spur eukaryotic diversification toward the end of the Proterozoic Eon.

If the rise of eukaryotes to ecological and taxonomic prominence came long after the origin of eukaryotic cells, we have to think about triggers for diversification, either biological or physical. What about sex? The undeniable (and, perhaps, not wholly scientific) attraction of this proposition rests on a simple observation and a bit of arithmetic. The observation is taxonomic: about 4,000 bacteria have been given species names; in contrast at least 100,000 protozoans and algae, another 100,000 fungi, some 300,000 land plants, and more than a million animals have been described. The math comes in to bolster the argument that sex promotes diversity by enabling eukaryotes to generate the raw material for evolution—genetic variation within a population—differentially well. Bill Schopf is especially fond of this idea, which he illustrates by means of a simple thought experiment. In bacteria that reproduce by binary cell division, ten mutations that arise in a population will result in a maximum of eleven different combinations of genes. But, let those same ten mutations arise in a sexually reproducing population of eukaryotes, and the possible genetic combinations run into the thousands. No wonder eukaryotes are so diverse. The argument is comfortably simple, it makes intuitive sense, and it relies on natural history of the kind that "every schoolboy knows." In short, it is just the sort of proposition that graduate students love to dismantle.

Let's start with the statistics on species diversity. Named bacterial species have one trait in common—they can be grown in the laboratory. Using new molecular techniques, however, microbial ecologists have found that culturable organisms represent as little as 1 percent of the bacterial diversity present in natural environments. Thus, the true diversity of bacteria may rival that of protozoans and algae.

We also have to poke a bit at the issue of genetic variability, because it makes the important assumption that bacteria have no means of combining genes from two individuals. As noted already in chapter 2, this assumption is spectacularly incorrect. The well-known intestine dweller *E. coli*, for example, practices a form of sex in which two cells become tethered by a tiny tube, allowing genetic material to pass from one to the other. Other bacteria take up small lengths of DNA released to the environment by dying cells. Still others incorporate genetic material transported by viruses. In fact, bacteria exchange genetic material all the time, and they do it not only between two individuals in the same population but between distinct species and even between kingdoms. If sex is defined as the exchange of genetic material between individuals, bacteria are decidedly sexy. Prokaryotes do not suffer from a poverty of genetic variation.

Thus, if we want to entertain the notion of sex as a trigger for eukaryotic diversification, we have to posit that early eukaryotes lacked both sex and the mechanisms for genetic exchange found in bacteria. At present, however, we don't know this. We really don't know when sexual reproduction entered the eukaryotic life cycle, if in fact it wasn't already present in the last common ancestor of all living protists. Perhaps we should approach the problem differently.

The "just add sex" hypothesis assumes that diversity is somehow constrained by the rate at which genetic variation can be generated, but diversity depends as much if not more on biological function and ecology. Prokaryotic diversity reflects the remarkable ability of bacteria and archaea to exploit specific nutrient sources and energy gradients. In contrast, eukaryotes diversified by approaching the world in new ways. As discussed in the preceding chapter, the eukaryotic cytoskeleton and membrane system enables nucleated cells to do something that bacteria can't do at all—swallow particles, including other cells. Thus, as evidenced by the vase-shaped microfossils in the Grand Canyon, eukaryotes introduced grazing and predation into microbial ecosystems. Doggerel by Jonathan Swift captures the consequences nicely:

> So, naturalists observe, a flea
> Has smaller fleas that on him prey;
> And these have smaller still to bite 'em;
> And so proceed ad infinitum.

By expanding ecosystem complexity, eukaryotic cells erected a new scaffolding for diversity.

Eukaryotic organisms do something else that has largely eluded prokaryotes. Plants, animals, fungi, and seaweeds develop via a complex pattern of cell division and differentiation choreographed by molecular signals that pass from cell to cell, switching specific genes on or off as they go. This beautifully coordinated regulatory system may have its origins in single-celled organisms that changed size or shape as they passed through the cell cycle, but it eventually made possible the evolution of complex multicellularity, and in so doing fueled a further expansion of eukaryotic diversity. Ninety-five percent or more of extant eukaryotic species are multicellular.

At the end of the day, it isn't necessary (or, perhaps, wise) to focus too narrowly on any one trait as key to eukaryotic diversification. Sex, cytoskeletons, genetic regulation, and, doubtless, other characters interacted to produce the plethora of eukaryotic forms seen today. No one knows when the modern eukaryotic "tool kit" was assembled, but the small fossils in Roper shales suggest that it was in place long before the diversification recorded in late Proterozoic rocks.

This brings us to the remaining explanation for late Proterozoic diversification—changing environments. Can we envision environmental changes that might have improved the odds of eukaryotic success in a prokaryotic world? If we can, do the rocks provide evidence that the required changes actually took place in concert with eukaryotic diversification? Increasingly, the answer to both questions appears to be yes.

In chapter 6, we discussed evidence for an increase in atmospheric (and surface ocean) oxygen 2.2–2.4 billion years ago. When I was a student in the 1970s, this event was generally accepted as the transition between Earth's two great environmental states: the Archean to earliest Proterozoic, when O_2 was scarce, and the past 2 billion years, during which air and ocean have both been bathed in oxygen. But, as noted in chapter 6, Odense University's Don Canfield has proposed that what the early Proterozoic oxygen revolution ushered in was not the modern world, but rather an alien intermediate marked by moderate oxygen in the atmosphere and surface sea and hydrogen sulfide in deep waters (figure 9.6).

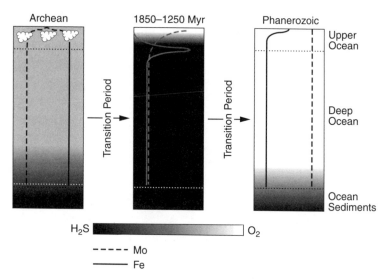

Figure 9.6. Triptych illustrating the three phases of ocean evolution. Early oceans contained little oxygen, but relatively abundant iron. Modern oceans contain abundant oxygen and little iron. In between, during a long-lived state that may have lasted from 1.8 billion years ago until near the end of the Proterozoic Eon, the oceans are thought to have had moderate oxygen in surface waters, but hydrogen sulfide at depth. In such an ocean, biologically important trace elements such as iron and molybdenum (concentrations, higher to the right, illustrated by vertical lines) may have been in seriously short supply. (Reprinted with permission from A. D. Anbar and A. H. Knoll, 2002. Proterozoic ocean chemistry and evolution: a bioinorganic bridge? *Science* 297: 1137–1142. Copyright 2002 American Association for the Advancement of Science)

Those fossiliferous rocks from northern Australia provide a good test of Don's hypothesis (maximizing my scientific return per tick bite). Putting our heads together, postdoctoral fellow Yanan Shen, Don, and I have been able to show that shales deposited in deeper parts of the Roper (and two older) basins preserve chemical signatures similar to those observed today in sediments at the bottom of the Black Sea. The Black Sea is well known to Earth scientists because its oxygen-rich surface layer blankets a much larger volume of water that fairly bristles with H_2S. Independent work by Tim Lyons at the University of Missouri and my former student Linda Kah, now at the University of Tennessee, corroborates and expands this view of distinctive mid-Proterozoic oceans.

Why should this matter to evolving eukaryotes? Ariel Anbar, a talented geochemist at the University of Rochester, pointed out to me that in seawater like that envisioned for the mid-Proterozoic, the essential nutrient nitrogen would be relatively scarce.[5] That is okay for cyanobacteria because they can fix nitrogen and, furthermore, are remarkably adept at scavenging biologically useful forms of nitrogen from their surroundings. But the situation is different for photosynthetic eukaryotes. Algae thrive today where levels of nitrate (NO_3^-, a biologically usable form of nitrogen that is readily available in modern seawater) exceed short-term requirements for growth, allowing cells to stockpile nutrients as a hedge against leaner times. In the mid-Proterozoic ocean, however, this would have been difficult, because nitrate can't build to high levels when oxygen is limited and sulfide lurks beneath the surface ocean. Algae can't fix nitrogen, and they compete poorly against cyanobacteria for the scarce nitrogen compounds that would have been present in mid-Proterozoic seawater. Moreover, to make use of such nitrate as might have been present, algae would have needed the metallic element molybdenum.

"Moly" is an essential ingredient in nitrate reductase, the enzyme that makes nitrate useful to organisms. Today, moly is widely available as a trace constituent of seawater, but in the alien world of the "Canfield" ocean, this element would have behaved somewhat differently. Then as now, moly would have been weathered from continental rocks and carried into the sea by oxygenated rivers. Unlike its distribution today, however, moly would have been common *only* in coastal waters where the rivers entered the sea. Farther from shore, surface waters would have mixed with subsurface water masses, causing molybdenum to be removed by reaction with hydrogen sulfide. Algae that washed into the open ocean would, thus, have faced the insult of molybdenum deprivation to go with the injury of nitrogen limitation.

All this suggests that 1.5 billion years ago, life might have been tough for eukaryotic algae. (Protozoans should have fared better, because they obtain the nitrogen they need from the cells they eat. But remember that

[5] Recall from chapter 2 that nitrogen gas (N_2) permeates air and ocean waters, but cannot be used directly by most organisms. Many prokaryotic microorganisms, including cyanobacteria, can "fix" nitrogen, converting gaseous N_2 into ammonium ion (NH_4^-) that can be incorporated into biological molecules.

the paleontological record provides few recognizable glimpses of early protozoans—with the conspicuous exception of those vase-shaped fossils in Grand Canyon rocks, the fossils that document late Proterozoic eukaryotic expansion are largely algal.) Returning once more to rocks of the Roper Group, we can ask how presumed algae were distributed in this mid-Proterozoic seaway. Our prediction is that photosynthetic eukaryotes should have been most abundant and diverse along the ancient seacoast, where nitrate levels would have been highest and molybdenum most readily available. That, it turns out, is just what we see.

Indeed, it appears that, globally, eukaryotic algae first took root in coastal waters and only later spread across continental shelves. This ecological expansion is poorly documented, but in general it appears to have commenced about 1.2 billion years ago as nitrogen limitation in the "Canfield" ocean began to weaken. Granted new ecological opportunities, both seaweeds and plankton diversified—as we see in the fossil record. Protozoans like those buried deep within the Grand Canyon must have diversified, too, as heterotrophs learned to exploit the new *biological* environments created by the algae.

Thus, as our view of Proterozoic geology and paleontology strengthens, it once again implicates environmental history in the determination of evolutionary pattern. That doesn't let biological innovation off the hook—new functional possibilities, especially those presented by multicellularity, must also have stoked the flame of eukaryotic diversification—but it requires us to seek evolutionary explanation in the *interaction* between genetic possibility and environmental opportunity.

The transition to a fully oxic world appears to have been protracted. As discussed more fully in chapter 11, oceans rich in oxygen from top to bottom may not have developed until the Proterozoic Eon was almost over. When they came, however, the culmination of Earth's environmental transformation paved the way for one last revolution in biology—the rise of animals.

10 | Animals Take the Stage

In latest Proterozoic rocks, we find at last what Charles Darwin predicted long ago—the fossilized impressions of early animals. But the fossils are not at all what Darwin expected. The ancestors of modern animals undoubtedly lived in latest Proterozoic seaways, but most end-Proterozoic fossils have unusual forms that separate them from, rather than link them to, Cambrian and younger faunas.

For paleontologists weaned on arctic research, the sun sets early in Namibia. By 6:00 P.M., packs and hammers must be stowed, and if we're not to stumble around by flashlight, firewood must be gathered. Temperatures may have reached 100°F in the afternoon, but by morning our sleeping bags will be rimmed by frost. As daylight wanes, the stark hills and fantastic shrubs of southwestern Africa fade from view. But even as they disappear, a new wonder takes shape in the evening sky— the Milky Way seen through clear desert air. Untold millions of stars form a broad arc across the southern sky, stars so densely packed that aboriginal Australians found their constellations in patches of emptiness amid the glimmer, rather than by connecting sparse points of light. Every few minutes, a meteor streaks across our celestial canopy.

The stars make good companions as we drift toward sleep, tightly wrapped against the cooling night air. But sleep can be fitful in the desert. The stars may disappear behind a curtain of clouds, prompting the worry that we'll soon be wet as well as cold. Or, zebras may amble past camp, their quiet hoof-falls nudging weary geologists awake. Other animals are less respectful. More than once, I've been awakened by piercing shrieks just beyond our fading campfire—a troop of baboons, irritated by human interlopers. Eventually, however, sleep returns until an amber rim on the eastern horizon heralds the end of stars and cold, and the start of a new working day.

The fossils of Gunflint and Spitsbergen offer few clues to the origins of creatures like zebras and baboons. Even Doushantuo phosphorites, formed just 50 million years before the Cambrian cliffs in Siberia, contain only microscopic hints of gathering animal evolution. But here in Namibian rocks deposited at the very end of the Proterozoic Eon (figure 10.1), and in beds of comparable age around the world, we see at last the palpable fuse of Cambrian Explosion—large animals that plausibly include the ancestors of our familiar biota. In some ways, it marks the realization of Darwin's dream—animal life before the Cambrian. But Namibian fossils also deepen Darwin's dilemma, because their unusual shapes challenge our efforts to locate them on the Tree of Life. Do these remains really trace a path to modern animals, or are they a dead-end fork on the evolutionary road?

Figure 10.1. Sedimentary rocks of the Nama Group rise out of the Namibian desert. The large gray mounds to the left of the mesa and near its top are microbial reefs that contain calcified animal fossils. Ediacaran impressions occur in sandstones that form the conspicuous ledge high on the hill.

I first visited Namibia more than twenty years ago, in the company of South African geologist Gerard Germs. A gentle man of philosophical bent, Germs emigrated from the Netherlands as a graduate student and took up the challenge of bringing geologic order to a poorly known region the size of Texas. His success laid the foundation for continuing research that has changed the way we think about early animal evolution. Much of that research has been directed by MIT's John Grotzinger, architect of our modern understanding of these rocks and their paleontological riches. John's driving desire is to understand how sedimentary rocks accumulate and, especially, to know how they did so on the early Earth, when life and environments differed from today. To address these issues, he must find places where thick successions of ancient rocks are exceptionally well preserved and unusually well exposed. Southern Namibia fits the bill perfectly: its Proterozoic sediments remain almost untouched by tectonics or metamorphism, but they are dissected by canyons that allow geologists to map stratigraphic relationships in three-dimensional detail. With his students, John has taken the latest Proterozoic rocks of Namibia apart and put them back together, in the process learning how tectonics, sea level, climate, and biology shaped the sedimentary record seen today. Along the way, he has discovered a host of new fossils, including large reefs, built by microorganisms but bristling with the skeletons of early animals. It was the opportunity to study those skeletons that lured me back to Namibia.

Sedimentary rocks of the Nama Group accumulated in a broad basin formed in response to continental collisions that forged the supercontinent Gondwana. In its lowermost part, the Nama succession consists of pebbly and conglomeratic sandstones formed on an ancient coastal plain. Above these are finer-grained sandstones deposited along an ancient coastline, followed by siltstones and shales deposited farther offshore. In the basin center, beyond the reach of silt and mud, limestones precipitated from clear waters. Light green ash beds, introduced by nearby volcanoes, preserve a record of time: deposition began about 550 million years ago and continued until the very end of the Proterozoic (543 million years ago), when uplift and erosion carved deep canyons into underlying rocks. These paleocanyons are filled by more sandstones and shales; diverse trace fossils and a 539 ± 1–million-year-old ash bed indicate that when Nama sedimentation resumed, the Cambrian Period was already under way.

The conventional hallmarks of Proterozoic biology, seen before from Spitsbergen to Siberia, appear once more in these Namibian rocks. Stromatolites are conspicuous if not abundant features of Nama limestones, and in Nama shales, cyanobacterial filaments lie buried with simple algal microfossils. Here, then, just below the Cambrian boundary, the paleontology still looks . . . well, Proterozoic. But, there is a difference. If we examine Nama sandstones carefully—preferably in late afternoon when the sun, set low in the sky, throws surface features into high relief—we see fossils that are almost shockingly different from anything found in older rocks (plate 7). We see the impressions of large, complicated organisms, as well as simple tracks and trails unambiguously made by animals. In truth, the Nama fossils are shocking whether we approach them from above or below, for if they have no counterparts in older beds, Nama impressions bear equally little resemblance to most fossils found in Cambrian or younger rocks. Thus, the debate: do the remarkable fossils in Nama and other latest Proterozoic rocks record the ancestors of modern animals or a failed evolutionary experiment at the dawn of animal evolution?

Fossils were discovered in Nama rocks as early as 1908, and between 1929 and 1933 the German paleontologist Gürich provided detailed descriptions of several species. Not much was made of this discovery, however, perhaps because the biological and geological frameworks needed to understand its importance were not yet in place. Scientists didn't fully appreciate either the genealogical relationships we know from the Tree of Life or the time relationships among ancient rocks. By 1946, however, when Reg Sprigg began to uncover similar assemblages in the remote Ediacara Hills of South Australia, the necessary frameworks had begun to take shape. Moreover, the Australian fossils found a worthy champion in the great paleontologist Martin Glaessner (along with Barghoorn, Cloud, and Timofeev, the fourth patriarch of Precambrian paleobiology). "Ediacaran" fossils, as they came to be known, were interpreted by Glaessner—and many who followed—as the exposed roots of the metazoan tree: the earliest representatives of animal phyla that blossomed into diversity in the ensuing Cambrian. Glaessner took an active interest in the Ediacaran fossils of Namibia, but it was discoveries in the 1970s by Gerard Germs and Hans Pflug, of Giessen University in Germany, that rekindled paleontological interest in Nama rocks. Another German scientist stirred the pot, as well. In 1984, Adolf Seilacher, one of the world's most distinguished paleontologists, announced that Glaessner had gotten it all wrong.

One more time we must ask how paleontologists interpret fossils. How, in this particular case, do they coax biology from Ediacaran impressions, casts, and molds? Evidence of anatomy or physiology was stripped away shortly after the carcasses were impressed into surrounding beds, leaving only form to guide our interpretations. Of course, morphology is *usually* all that remains for paleontologists to ponder. Dinosaurs left only their bones (and, rarely, traces of skin), but that's enough to reveal volumes about their biology, because dinosaur backbones, ribs, and teeth display morphological landmarks that relate them unambiguously to vertebrate animals alive today. The same is true for trilobites; they may be long extinct, but their segmented bodies and jointed legs tie them to living horseshoe crabs, shrimp, and other arthropods. Therein lies the problem. The impressions in Namibian sandstones have unfamiliar shapes, making it difficult if not impossible to map their features onto the forms of living animals.

The simplest impressions in Nama rocks are shallow disks up to a few inches across—the kind of fossils one might expect to be formed by jellyfish buried in a storm (plate 7b). Indeed, for many years jellyfish were considered likely counterparts of these fossils, but this interpretation has a serious flaw. Disklike fossils are common in Ediacaran-age sandstones, and nearly all occur as casts that bulge downward from the *bottoms* of sandstone beds. In other words, they are casts of organisms that formed *depressions* in the shallow seafloor. For jellyfish to form such fossils, they would have to land upsidedown (and with considerable impact) when cast onto sediment surfaces. We need only stroll along a beach in the wake of a storm to convince ourselves that this is not how jellyfish land. More likely, the disklike fossils lived on the seafloor, nestled into sediments much like modern sea anemones, bottom-dwelling cousins of the jellyfish. Other discoidal fossils represent originally bulbous or conical holdfasts, or anchors, that tethered more complicated constructions to the seafloor. And still others may not be animals at all—common ball-shaped fossils called *Beltanelliformis* appear to have been fluid-filled seaweeds (plate 7d) comparable in size and structure to the living green alga *Derbesia*.

Relatively few disklike fossils have been found in Namibia, but elsewhere—in Ediacaran rocks from Australia and spectacularly fossiliferous beds from the White Sea region, Russia—these rounded fossils are far and away the most common components of Ediacaran faunas. There is *Cyclomedusa*, up to five inches across and marked by concentric folds and radial grooves—like a striated cone collapsed along its axis. Then there is *Mawsonites* (plate 7b), similar in size but ornamented by concentrically arranged lobes or bosses. *Medusinites* is smaller (less than two inches across) and smoother, but has a sharply defined circular groove in its center. In contrast, *Ovatoscutum* sports closely spaced, parallel grooves, as though it were fashionably decked out in corduroy. Thin tentacle-like projections ornament the disk of *Hiemalora*. And the lobed *Inaria* resembles a head of garlic flattened against the seafloor.

Most paleontolgists agree that, in general, these disks represent anatomically simple, bottom-dwelling animals related to living Cnidaria, the animal phylum that includes sea anemones and jellyfish. Even those disks shown to be the anchors of colonial animals may find their genealogical home within or near the Cnidaria, as discussed below. Perhaps the most unusual interpretation of Ediacaran disks is that proposed by Dolf Seilacher, the German challenger to Glaessner. Seilacher suggested that least some of these fossils preserve bizarre "sand corals" that ingested sediment as a sandy ballast to hold them in place—"a rock in a sock," in Dolf's clever phrase. Others, however, including me, believe these fossils are merely "socks in a rock"—conventional organisms cast in sand.

While the closest living relatives of Ediacaran disks may occur among the Cnidaria, no one suggests that Proterozoic disk formers were identical to species alive today. They are extinct taxa that record an early radiation of anatomically simple animals.

A second group of Ediacaran fossils consists of complex, often leaflike, forms made up of repeating tubular units (plate 7a, c, and e). Called vendobionts, these fossils are represented in Namibia by several species. *Rangea* is an elongate fossil, up to about six inches long, consisting of a central axis from which two rows of branches emerge; the branches contain numerous interlocking tubes, each a few millimeters wide. This distinctive form reminds many paleontologists of sea pens, modern relatives of jellyfish and sea anemones in which simple individuals form complex, leaflike colonies. If *Rangea* is a colonial animal, then the repeating tubular units may represent its constituent individuals.

Sea pen analogies become harder to sustain, however, when applied to other Nama vendobionts. *Pteridinium*, a second leaflike form found abundantly in a mid-Nama storm bed, superficially resembles a sea pen, but it has three "wings" instead of two, and each wing contains a single rank of tubes arranged perpendicular to the main axis (plate 7e). No modern sea pen looks like that! *Swartpuntia*, found by John Grotzinger at the very top of the Proterozoic succession, also has its tubes arranged in three broad wings attached to a stout central axis— like a Chinese fan imagined by Picasso (plate 7a). And *Ernietta* is even worse—its many elongate tubes form a complex cup, possibly open at the top.

Vendobionts provide a Rorschach test for paleontologists. Individual fossils have been interpreted as colonial cnidarians, as segmented worms, as primitive arthropods, seaweeds, lichens, and more. At the same time, some paleontologists insist that all vendobionts share a common architecture and so, a common ancestry. It is here that Dolf Seilacher mounted his most audacious challenge. Dolf proposed, not only that vendobionts were all cut from the same cloth, but that the cloth no longer exists. Unimpressed by Glaessner's interpretations of Ediacaran fossils as early offshoots of phyla seen today, Dolf declared that vendobionts were quiltlike organisms made up of cylindrical tubes filled with "plasmodial fluid" rather than cellular tissue. Interpreted this way, vendobionts seem foreign—more alien than animal—and this was precisely Dolf's point. In 1992, he formally proposed the Vendobionta as an extinct kingdom separate from the animals, an experiment in macroscopic multicellularity that blossomed for a geological moment but ultimately failed. Earlier interpretations of vendobionts raised only eyebrows; Dolf's raised hackles, and so inspired a new generation of research.

Nama fossils serve to introduce the odd morphologies of vendobionts, but other localities show their true diversity. *Charniodiscus* is a spectacular fossil first discovered in Australia. Like *Rangea*, it is leaflike in form, with a large (three-inch) disklike holdfast that anchored an originally erect axis to the seafloor. Thirty to fifty lateral branches depart in rows from either side of the central axis, and each branch bears a flap on one face marked by parallel grooves. The entire fossil can be more than a foot long. In Seilacher's view, *Charniodiscus* is a typical vendo-biontid alien. But Richard Jenkins, a respected paleontologist at Ade-laide University, interprets *Charniodiscus* more conventionally as a colo-nial cnidarian organized much like a living sea pen. (Unlike living sea pens, however, the branches of *Charniodiscus* were fused together to form an unbroken surface.)

Charnia, first discovered in the Charnwood Forest of England and now also known from Newfoundland, Australia, and Russia, superfi-cially resembles *Charniodiscus*, but has no central axis; its branches con-sist of tubelike units arranged in parallel and tightly joined between rows to form an intricately quilted surface. *Phyllozoon* was also quilted, but apparently spread across the seafloor like a miniature throw rug.

Perhaps the most thoroughly studied—and, therefore, most hotly contested—vendobiont is *Dickinsonia*, known from large populations in Australia and the White Sea. *Dickinsonia* (plate 7c) is an elliptical fossil made up of cylindrical tubes joined along their long axes to form a con-tinuous surface; specimens may be as small as a penny or as large as a turkey platter, but they are never more than a few millimeters thick. A narrow but pronounced ridge runs down the middle of the long axis. Mary Wade, a colleague of Glaessner's at the South Australian Mu-seum, first described *Dickinsonia* as an annelid worm, interpreting the transverse tubes as body segments and the central ridge as a gut cavity. She proposed the living annelid worm *Sphincter* as a modern counter-part, although the flattened shape of this worm is highly unusual (and not at all primitive) within the phylum. In contrast, Misha Fedonkin, the affable Russian who brought the White Sea fossils into focus, proposed that the tubular segments of *Dickinsonia* meet along the central axis but do not cross it. If Fedonkin is correct—and more than one disciple of Glaessner has contested his interpretation—*Dickinsonia* could not pos-sibly be an annelid. To Dolf Seilacher, of course, *Dickinsonia* is just one more extinct vendobiont.

Was *Dickinsonia* a worm or a failed experiment, an early ancestor of familiar animals or an extinct life-form only distantly related to living invertebrates? Interpretation isn't easy, and I can easily find someone to disagree with any opinion I might venture. But a few clues shed light on the case. UCLA's Bruce Runnegar and Jim Gehling of the South Australian Museum discovered folded specimens, which show that *Dickinsonia* had a flexible body. A handful of fossils also provide evidence that tubes could contract and so must have contained muscle cells. And rare specimens in which tubes ripped open but retained their cylindrical shapes suggest that whatever filled these structures, it wasn't "plasmodial fluid." That's one strike against the vendobiont hypothesis. On the other hand, *Dickinsonia* shows no evidence of the organ systems expected in annelid worms: there is no mouth at the end of the medial ridge, no hairlike setae, and no parapodia (stumpy leglike appendages on the body segments of marine annelids). Early in the game, it was possible to blame such absences on preservation, but that won't work anymore. What we see is pretty much what was there.

Ediacaran rocks contain trackways that could have been plowed by smaller dickinsonids, but few trails that could, even in principle, be associated with larger specimens. Apparently, then, *Dickinsonia* reclined on the sediment surface but did not move across it, very unlike a worm. And one more observation supports the interpretation of *Dickinsonia* as a vendobiont—the arrangement of its tubes closely resembles that seen in the fanlike wings of the three-pronged Nama fossil *Swartpuntia*.

I confess to some uncertainty in interpreting these perplexing fossils. Vendobionts don't seem to develop like modern algae and they don't look like living worms. But neither am I convinced that they record, like Shelley's statue of Ozymandias, a vanished kingdom in an antique land. What is the alternative? Taking a cue from *Rangea* and *Charniodiscus*, I suspect that most vendobionts were colonial animals at least broadly related to the living Cnidaria. Today, coloniality is widespread among cnidarians, from the Portuguese man-of-war that floats on the sea surface (its float, stinging tentacles, and reproductive structures are all anatomically complete individuals) to the massive reef corals and delicate sea fans that proliferate on the ocean floor. In the absence of well-developed organ systems, cnidarians achieved complexity by differentiating *individuals* within colonies, and this may have been the case for vendobionts, as well.

Tubelike individuals in vendobiont colonies must have had a simple anatomy, but, like living cnidarians, they could have had both nerve nets and functionally coordinated muscle cells. Moreover, the structurally competent material within tubes could have been an inert substance like the "jelly" in jellyfish. And, as Mark McMennamin of Mount Holyoke College first speculated, vendobionts might have gained nutrition from symbiotic algae or bacteria, as (once again) do many living cnidarians. "Could have," "might have"—much remains tantalizingly beyond our grasp. However, by viewing the complex shapes of vendobionts as colonies built by individually simple animals, we can interpret most if not all of these fossils as members of a single clade, and one that vanished long ago. I suspect that the vendobionts do not comprise an extinct parallel to the animal kingdom, but rather record early animals that possessed some but not all of the features found in living cnidarians. (Cnidarian features *not* found in vendobionts include a mouth fringed by tentacles.) Indeed, in limited retreat from the extinct kingdom hypothesis, Dolf Seilacher and Yale's Leo Buss have suggested such a possibility.

Sherlock Holmes famously recognized the importance of absence—of missing clues and things that did not happen:

> "Is there any other point to which you would wish to draw my attention?"
> "To the curious incident of the dog in the night-time."
> "The dog did nothing in the night-time."
> "That was the curious incident," remarked Sherlock Holmes.

Absence is also worth noting in Nama and other latest Proterozoic rocks, and the fossils most obviously missing are those found so conspicuously in Cambrian rocks deposited only 10–20 million years later. Ediacaran disks and vendobionts suggest that cnidarian-like animals were common in latest Proterozoic ecosystems, but where were the ancestors of trilobites, of mollusks, of brachiopods, . . . of *us*?

Absence of evidence is suggestive, but when are we justified in interpreting it as evidence of absence? The answer, as always, lies in sampling. Cambrian rocks tell us where and under what conditions complex animals became fossilized, and only when we have searched thoroughly in all the right Proterozoic rocks can we have confidence in an evolutionary interpretation of missing fossils.

Cambrian rocks contain the tracks, trails, and burrows of anatomically and behaviorally complex animals that inhabited shallow seaways; in the cliffs along the Kotuikan River, their abundance and diversity increased dramatically as we climbed higher into the Cambrian succession. Nama sandstones also contain tracks and trails made by early animals as they moved through bacteria-rich sediments just below the surface of the seafloor. The trackways are small and simple—about the size and shape of a spaghetti strand dropped carelessly onto the floor (figure 10.2). The tracks run parallel to the sediment surface and rarely penetrate to even modest depths. But, based on observations of living trail-formers, it appears that most of these traces were made by creatures more complicated than sea anemones and jellyfish. Animals that make tracks like these have one distinctive feature in common—bilaterally symmetric bodies in which a single plane of symmetry runs from head to hind end. *Bilaterian* animals (discussed further in chapter 11) include all anatomically complex metazoans from trilobites to vertebrates. Trace fossils document their presence in latest Proterozoic oceans. Might bilaterian animals also lurk among Ediacaran impressions?

Figure 10.2. Simple track fossils made by bilaterian animals in a latest Proterozoic seaway. The specimen comes from South Australia. Each trail is a bit over 1 millimeter wide.

Some fossils from Australia and the White Sea are neither disks nor vendobionts, and these provide a hunting ground for more complex animals. *Tribrachidium*, for example, is a circular cast that might easily be lumped with Ediacaran disks, except that it displays three large grooves that branch repeatedly as they spiral outward from its center. Along with a handful of related fossils from the White Sea, *Tribrachidium* has variously been allied with the sponges, cnidarians, or echinoderms. Its affinities, however, remain problematic—three-part symmetry is rare among living animals. Functionally, the canal-like internal structure suggests an animal that flushed large volumes of seawater through its body, as sponges do today.

More promising are other small fossils that resemble arthropods to a greater or lesser degree. *Parvancorina* consists of shieldlike molds, mostly less than half an inch long (figure 10.3). A pronounced rim runs along the outer margin, while the interior displays a T-shaped ridge whose top curves along the rounded (front?) edge; faint lines sometimes interpreted as legs can be seen beneath the shield. *Parvancorina* certainly exhibits bilateral symmetry, and looking at it, one can hardly avoid thoughts of trilobites. But, of course, *Parvancorina* is not a trilobite,

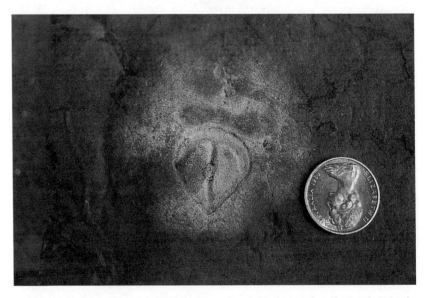

Figure 10.3. *Parvancorina*, a problematic fossil that superficially (and, I think, only superficially) resembles trilobites.

and though it has certain features reminiscent of this group, it lacks a host of additional characters evident in trilobites, lobsters, crabs, and all other arthropods.

The same can be said about an assortment of other Ediacaran fossils, including *Praecambridium* and *Vendia. Spriggina,* named for Ediacara pioneer Reg Sprigg, is particularly intriguing in this regard; its two-inch body has both segments and what appears to be a rounded head shield—very like an arthropod. Dolf Seilacher rather hopefully interpreted it as a vendobiont made up of interlocking tubes, but many other paleontologists see *Spriggina* as a segmented, bilaterally symmetric animal, "arthropoid" if not truly arthropod.

Finally, there is *Kimberella,* a small organism that looks like a smoked mussel preserved in rock. According to Misha Fedonkin and Ben Waggoner, of the University of Central Arkansas, this resemblance is more than coincidental. They interpret *Kimberella* as a bilaterally symmetric organism that possessed a muscular foot for locomotion, a baglike body packed with visceral organs, and a tough organic mantle over its back. These features all occur in mollusks, the phylum that includes mussels and clams, along with snails and squid. Like *Spriggina,* however, *Kimberella* lacks other features found in the living animals to which it is compared.

Such fossils are both exciting and frustrating—exciting because they show flashes of a familiar biology and frustrating because individually familiar characters occur in decidedly unfamiliar combinations. If, however, we can lay aside the urge to see modern animals in Proterozoic fossils and accept Ediacaran morphologies for what they are, the balance tips strongly toward excitement. The unmistakable sense I get from Ediacaran fossils like *Kimberella* and *Spriggina* is that they preserve early stages in the assembly of features that collectively mark Cambrian animals as "modern." Ediacaran species weren't there yet, but at least some were on their way.

Contrasting the simple trace fossils below the Proterozoic-Cambrian boundary with the more abundant, diverse, and complicated tracks and burrows found above supports the idea that something big really did happen at the start of the Cambrian Period. But much of our sense of Cambrian diversity comes from skeletons preserved in carbonate rocks. The Nama Group contains plenty of limestones; thus, it offers one more test of evolutionary pattern. Do Nama carbonates contain ancient skeletons? If so, do these fossils establish biological continuity between Proterozoic and Cambrian animals, or do they reinforce the differences between the two?

Generations of paleontologists regarded the origin of mineralized skeletons as synonymous with the Cambrian Explosion. In 1972, however, Gerard Germs showed this view to be wrong. In the course of his field research in Namibia, Germs discovered small tubes made of calcium carbonate in Nama limestones (Figure 10.4a). Germs christened these fossils *Cloudina*, in honor of Preston Cloud, and recognized two species, one of which he named for his mother. *Cloudina hartmannae* and *C. reimkeae* differ in size—the first an inch or two long and a quarter inch wide; the second smaller by a factor of two—but they share a common organization. The fossils are gently curving cylinders ornamented by closely but irregularly spaced flanges that protrude outward—the whole resembling a series of tiny funnels stacked one inside the next. Pogonophoran worms (distant relatives of earthworms) live in tubes built like this, but simpler animals can form mineralized tubes, as well. In fact, rare specimens found in China are branched, suggesting that, like Ediacaran disks, *Cloudina* might be related to sea anemones and jellyfish. Skeleton walls are thin and appear to have been flexible in life— *Cloudina* probably sported only a light coat of calcium carbonate.

Cloudina is important because it shows that animals learned how to build mineralized skeletons well before the Cambrian Period began. But how much do we wish to make of this? Is *Cloudina* a singularity—the exception that proves the rule of Cambrian biomineralization? Or, was it part of a larger assembly of latest Proterozoic animals that presaged the skeletal diversity of Cambrian faunas?

Here, at last, we return to the Nama reefs mapped by John Grotzinger. The reefs are astonishing to see in the field—great lumps as much as 200 feet high that rise out of the desert floor, excavated in recent millennia by erosion that peeled away encompassing shales (figure 10.1). The architects of the reefs were microorganisms, likely including algae as well as cyanobacteria, but skeleton-forming animals also found happy homes in small niches set high above the flat seafloor. Animal fossils are abundant in Nama reefs, and, judging from cross sections seen on weathered rock surfaces, they came in a variety of shapes and sizes (fig-

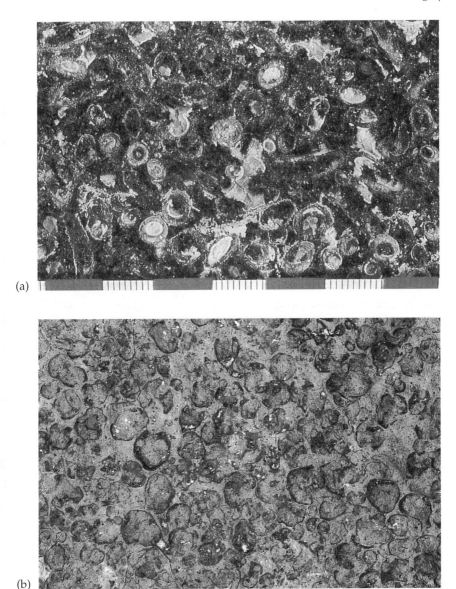

(a)

(b)

Figure 10.4. Calcified fossils from microbial reefs of the Nama Group. (a) *Cloudina*, tubular fossils lightly mineralized by calcium carbonate. (b) *Namacalathus* population showing the variety of shapes seen on rock surfaces in Nama limestones. Note centimeter scale bar in upper photo.

ure 10.4b). Tubes are common, although only a few display the flanges that identify them as *Cloudina*. More abundant are rounded cups up to an inch or so wide. Then there are goblet-shaped fossils with cup above and tube below, and fossils with clear hexagonal symmetry.

John and I spent hours crawling along these reefs, trying to figure out what kinds of animals were present and how many different species were represented. Those are difficult questions to address in the field, because the fossils can't be freed from the rocks that contain them. To find an answer, we had to cart large slabs of rock back to Cambridge, where John designed a system to prepare a smooth surface of each, and then slowly grind away the surface 25 microns at a time, taking a carefully registered digital photographic image after each shave. Using software originally developed for medical research, the library of digitized cross sections was assembled into three-dimensional virtual fossils (figure 10.5). Wes Watters, a bright MIT student with a head for physics but paleontology in his heart, did most of the work.

The computer models are eerily lifelike—they seem to bob and sway in some virtual current. The reconstructed fossils look like flexible wine glasses, with a cylindrical stem that opened upward into a round cup as much as an inch wide. Six (rarely, seven) regularly spaced holes impose a hexagonal pattern on the cups. As in *Cloudina*, the walls of these cups were thin and flexible, and so could have been mineralized only lightly. Evidently, early animals had little need for the robust mineral skeletons that protected their descendants from predators.

Given virtual fossils, we can simulate cross sections in any plane we wish. And when allowance is made for crushing and bending, nearly all of the tubes, cups, and goblets seen on reef surfaces can be interpreted as slices through a single form. Scyphopolyps, goblet-shaped relatives of (once again) jellyfish that attach to seaweeds in the present-day ocean, provide at least a general guide to the Nama fossils.

Clearly, *Cloudina* was not the only skeleton former in the Nama seaway. *Namacalathus*, the "goblet of Nama," thrived wherever microbial communities paved the seafloor, and continuing studies show that other mineralizing species were present as well, including coral-like animals that colonized cracks in the reefs. But just as clearly, diversity was limited. We may have discovered abundant new fossils in the reefs of Nama, but there are no bivalves or arthropods, no brachiopods or echin-

Figure 10.5. Virtual fossils of *Namacalathus*, reconstructed from digitized images, as described in the text. (Image generated by Wes Watters)

oderms. When we have searched everywhere we can, latest Proterozoic life still looks very different from the Cambrian.

On a sun-baked afternoon in Namibia, John Grotzinger and I stroll along the crest of an isolated hill, taking one more look at some of the youngest Proterozoic rocks exposed here or anywhere. The rocks are full of fossils: the vendobionts *Swartpuntia* and *Pteridinium*, calcareous *Cloudina* and *Namacalathus*, and trace fossils of modest diversity. These preserve a record of animals—but animals of distinctly Proterozoic aspect, far different from the diverse and complex invertebrates found in Kotuikan cliffs. The most conspicuous Nama fossils are to Cambrian animals what dinosaurs were to the mammals that graze on the plain below us—ecological antecedents but not direct ancestors.

Nama fossils would provide cold comfort to Darwin, who believed that Cambrian complexity took shape gradually over long stretches of Proterozoic time. Now it seems that the familiar biology of Cambrian seas emerged only as the Cambrian Period began. What was wrong with Darwin's solution, first considered along the Kotuikan River? How far *had* animal diversification progressed by the end of the Proterozoic Eon, and what ushered in the Cambrian world?

11 | Cambrian Redux

The complex forms of modern animals emerged only during the Cambrian Period, taking shape over a time span of at least 10 to 30 million years. Emerging insights into the genetics of development help us to understand the tempo and mode of Cambrian evolution, but we also need to factor in ecology—both the permissive ecology that enabled early variants to gain a foothold in the oceans and the ecological interactions among species that guided subsequent diversification.

We shall not cease from exploration
And the end of all our exploring
Will be to arrive where we started
And know the place for the first time.
   —T. S. Eliot
    Little Gidding

For the last time, we nose our rafts into the shore, eager to explore one final outcrop before the low drone of our approaching helicopter signals the end of another field season. Securing our boats to cobbles that line the stream, we take in the gray-gold cliffs that rise in front of us. The rocks look familiar, and they are. Having traveled across the globe and through 3 billion years of history, we have arrived once more at the Cambrian cliffs along the Kotuikan River. But now, as Eliot understood, we can see them with fresh eyes and so know them in new ways. Our peregrinations have, in fact, revealed the essential truth of Cambrian evolution: life has deep Precambrian roots, but the complex forms of Cambrian animals do not. There is nothing like the Cambrian until the Cambrian.

The Cambrian Explosion is the culmination of Precambrian evolution but a departure from it, as well. Can we build an understanding that captures both the continuity and revolution of Cambrian biology?

To understand where life stood as the Proterozoic ended, and how it changed during the ensuing Cambrian, we need a map, provided—as ever in evolutionary biology—by phylogeny. I noted in chapter 2 that anatomy and morphology early on revealed the relationships of some animal groups, but understanding how creatures as disparate as birds, clams, and tapeworms relate to one another challenged zoologists for many years. Embryology helped—for example, clams and polychaete worms share few features as adults, but many as larvae. But only in the age of molecular biology have the most difficult problems of animal phylogeny begun to yield. Figure 11.1 shows the animal tree as currently understood. We can climb it, using structure, function, and development as handholds and introducing fossils as we ascend.

Animals are not just overgrown protozoans, and they probably wouldn't have succeeded if they were. But what did metazoans do differently that enabled them to thrive on a crowded planet? Early branches of the tree highlight the achievements of multicellularity.

The closest known relatives of animals are the choanoflagellates, an

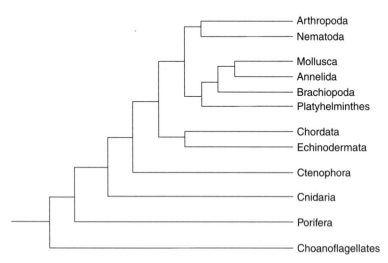

Figure 11.1. Evolutionary relationships among animal phyla, as indicated by molecular phylogeny.

unusual group of colony-forming protozoans. Choanoflagellate cells sport a distinctive collar around their flagella, rather like Dutch burghers in a Rembrandt painting. The presence of similar collars on the food-gathering cells of sponges has long implicated choanoflagellates in animal origins. But not all sponge cells share this feature, underscoring a distinct difference between choanoflagellates and animals. In animals, the many cells that arise from a single fertilized egg vary in form and function, allowing metazoans to "multitask" in a way that protozoans can't match.

Sponges don't simply produce different cell types; they array these differentiated cells into larger structures that facilitate food gathering and gas exchange. Sponges commonly grow into vases with hollow centers, porous walls, and an opening at the top. Collar cells that line the inner surface beat in unison, setting up water currents that bring in food particles and whisk away waste. A mosaic of flattened cells tiles the outer surface, while amoeba-like cells patrol the gelatinous zone in between, secreting fibrous proteins and, in some cases, mineralized skeletons of interlocking spicules.

Sponges undoubtedly diverged during the late Proterozoic, but they are not common in Ediacaran assemblages. Only with the evolution of mineralized skeletons near the Proterozoic-Cambrian boundary did sponges rise to paleontological prominence; skeletal fossils document dramatic Cambrian diversification within the phylum. Some early sponges secreted siliceous spicules—more than 90 percent of the five thousand sponge species found today fashion skeletons of silica, proteins (think of bath sponges), or both. Another group that persists to the present formed spicules, and sometimes massive skeletons, of calcium carbonate. Carbonate-secreting sponges called archaeocyathans were among the most diverse animals in Early Cambrian oceans, but mass extinction depleted their ranks halfway through the period, and for unknown reasons the group disappeared completely by the end of the interval. From ecosystem-dominant to evolutionary dead end in 20 million years—*sic transit gloria mundi*.

Before climbing higher in the tree, we need to address a fundamental question: how do multicellular organisms differentiate their exquisitely complicated bodies as they grow? Complex multicellularity requires cohesion among cells, communication between cells, and a genetic program to control cell differentiation during development. Cohesion keeps dividing cells from dispersing, making possible the precise spatial organization that underpins multicellular function. In seaweeds and plants, walls made of cellulose or other polysaccharides glue adjacent cells together. Animal cells, however, have no walls, so a battery of extracellular molecules must be deployed to make cells stick to one another; collagen, the protein that builds cartilage in humans, provides a prime example. Sponges produce a bevy of extracellular proteins that bind cells into vases; more complicated animals express similar proteins, but in greater variety.

In complex algae and land plants, thin strands of cytoplasm connect adjacent cells through tiny openings in their walls, providing a direct avenue for intercellular communication. In animals, molecular channels called gap junctions do much the same. Additionally, proteins embedded in the cell membrane bind chemical signals, setting off chain reactions that relay molecular messages to the nucleus. Cell surface proteins, thus, facilitate communication as well as cohesion, allowing groups of cells to function in a coordinated manner.

Communication is also key to animal development, the remarkable process by which fertilized eggs give rise to anatomically complex adults. When stressed, many single-celled eukaryotes cover themselves in a protective wall and suspend all but the most crucial cellular activities. That is, they differentiate into a distinct type of cell in response to signals from the environment. External signals also guide cell differentiation in animals, but in this case, the signals come from neighboring cells. A relatively small complement of genes, sometimes known as the developmental tool kit, coordinates the precise patterns of cell division, cell differentiation, and even programmed cell death that literally form hearts and minds. For the most part, these genes are not molecular carpenters assigned to build specific structures. They're middle managers that receive instructions from one gene and pass them to the next. Programmed development, therefore, proceeds according to complex networks of gene interactions that collectively regulate growth. The genetic tool kit of sponges is relatively simple, those of complex animals like flies or mammals more richly elaborated. Similar regulatory networks guide development in plants and algae, although many of the participating genes differ.

Sponges form one great limb of the animal tree; all other animals fall on the other.[1] More complicated animals, in turn, can also be divided into two major branches, the Cnidaria and the Bilateria (figure 11.1). We met these groups in chapter 10. Cnidarians comprise the jellyfish, corals, sea pens, and other taxa that provide structural analogues for many Ediacaran fossils; bilaterian animals, known mainly from trackways in Ediacaran sediments, today include an astonishing range of species from flatworms to whales.

As a group, cnidarians are distinctly more complicated than sponges— they have more types of cells, including muscle cells and a simple nerve network. Moreover, in cnidarians (and bilaterian animals), extracellular proteins bind cells into coherent sheets called epithelia that divide the animal body into compartments. Unlike sponges, therefore, cnidarians can form discrete tissues.

All cnidarians conform to a simple body plan—a hollow bowl or cylinder, with armlike tentacles around the opening (mouth). Two tissue layers that differentiate early in development line the inner and outer surfaces of the body, sandwiching gelatinous material in between (the "jelly" of jellyfish). The outer tissue, called *ectoderm*, contains muscle cells, nerves, and *cnidocytes*, specialized cells armed with tiny poison-tipped harpoons, coiled and ready for action. (If you have ever been stung by a jellyfish, you have firsthand experience of cnidocytes.) The inner *endoderm* bristles with cells that secrete digestive enzymes. Cnidarians do not build complex organs that integrate several tissues, like the heart or stomach of a mammal. However, as explained in chapter 10, they gained complexity in another way—by differentiating functionally specialized individuals within colonies.

---

[1] Some molecular phylogenies indicate that cnidarians and bilaterian animals are specific relatives of carbonate-precipitating sponges, with siliceous sponges alone on the first branch. Several features of cell ultrastructure support this view, although it remains a subject for debate. If correct, more complicated animals must have arisen from *within* early diverging sponges.

Collectively, muscle cells, nerves, and cnidocytes blazed a new trail in animal function. Sponges filter seawater to gather food particles, but cnidarians are predators, capturing prey with their harpoon-studded tentacles and stuffing it into their internal cavity for digestion. (As discussed earlier, reef corals and some other cnidarians have taken up farming, incorporating symbiotic algae into their tissues. Nonetheless, the Cnidaria fundamentally gather food by catching it.) Jellyfish and their relatives also *move*, further aiding the hunt.

Some protozoans catch and eat other cells, but with animals, predation took on an entirely new dimension. Protozoans might gobble cells singly or by the handful, but by filtering seawater with comblike organs, animals could catch them by the thousands and tens of thousands. And large size no longer provided safe haven. Animals, as well as microorganisms, had to avoid capture, and seaweeds had to cope with grazing. In effect, predatory animals became enormously important parts of the effective environment. The ensuing arms race between predator and prey has fueled evolution for more than 500 million years.

Many Ediacaran fossils probably relate to cnidarians, although most appear to represent early and extinct twigs on this branch. About ten thousand cnidarian species populate the modern oceans.

Remaining animal species—all 10 million of them,[2] including humans—belong to the Bilateria. Bilaterian animals differ from the Cnidaria in three fundamental ways. As explained in chapter 10, a single plane of symmetry divides the bilaterian body into left and right sides from head (more or less differentiated in most bilaterians) to tail. Moreover, three rather than two cell layers differentiate early in development—an ectoderm that contributes skin and nerve cells, an endoderm that gives rise to the digestive system, and an intervening layer called the mesoderm that differentiates into muscles and the reproductive system, among other things. Like cnidarians, bilaterian animals form tissues. Unlike cnidarians, however, bilaterians combine tissues

[2] The number of animal species alive today remains unknown. About 1.5 million species have been described (more than half of them insects), but the actual number could be much higher. Ten million lies near the midpoint of current estimates.

into complex organs, once again opening up new and diverse functional possibilities.

Cnidarians may have invented animal predation, but bilaterians perfected it. With organ systems came rapid swimming; muscular appendages to grasp and hold prey; mouths lined by mandibles, teeth, or rasping organs; sophisticated sensory organs including well-focused eyes; and, especially, brains able to coordinate the complex interactions of all these systems.

Increased predation intensified the need for protection. Some animals avoid predators by hiding. Others secrete poison. A third solution, discovered independently by many different groups, is armor—mineral-impregnated skeletons that protect against teeth and claws. *Cloudina* and *Namacalathus* show that at least a few late Proterozoic animals had lightly calcified coverings, but skeletons really took off in the Cambrian. The consequences for biology were significant, ratcheting up the evolutionary arms race and challenging predators to evolve structures that could pierce the defenses of prey. Mineralized skeletons also opened new functional pathways—for example, burrowing clams use their shells to dig into the sediment. Of course, the consequences for paleontology were also enormous. Skeletons preserve well in sedimentary rocks, increasing the likelihood that their makers will leave a fossil record. In fact, some geologists have argued that the Cambrian Explosion reflects the evolution of apparency—an explosion of fossils and not species. Such ideas, however, don't withstand close scrutiny. Calcified fossils are common in latest Proterozoic reefs, but they don't show any sign of the diverse and morphologically complex forms found in Cambrian and younger rocks (plate 8). Nor do late Proterozoic fossils replicated by calcium phosphate or compressed in shale even hint at the diversification to come. And, independently of any skeletons, trace fossils document a dramatic Cambrian diversification of animal behavior (figure 11.2). As emphasized by Stefan Bengtson, of the Swedish Museum of Natural History, skeletal evolution must be understood as part and parcel of the broader Cambrian diversification of animal life. Animals fashioned skeletons of calcium carbonate, calcium phosphate, silica, or just aggregated sediment particles—structural and biochemical innovations driven by the rise of sophisticated predators. (Of course, given the alternative strategies of speed, camouflage, and toxins, not all animals

Figure 11.2. U-shaped burrows made by invertebrates in Cambrian beach sands. Polychaete worms make similar burrows today.

invested in mineralized skeletons. Only about one-third of the modern marine fauna make preservable skeletons, and in Cambrian oceans the percentage may have been even lower.)

Bilaterian animals display a bewildering variety of form, but developmental and molecular data group them into three great clades (figure 11.1). Based on shared features of larval development, nineteenth-century zoologists united our own phylum, the Chordata, with echinoderms (starfish, sea urchins, and sea cucumbers) and a small phylum called the Hemichordata to form a supergroup dubbed the Deuterostomia. Molecular sequence comparisons support this evolutionary clustering of phyla and further divide the rest of the bilaterians (called Protostomia) into another pair of supergroups. Arthropods, nematodes, and several minor phyla link together as the Ecdysozoa—all animals in this clade form external cuticles and molt as they grow. The other great clade, infelicitously named Lophotrochozoa, includes such distinctive animals as mollusks, earthworms, brachiopods, and flatworms. Many but not all have spirally arranged cells in young embryos.

All living members of a phylum share a set of molecular and morphological features that reflects their descent (with modification) from a common ancestor. For example, insects, crustaceans, and centipedes look very different from one another, but all are variations on a single architectural theme based on segmented bodies, jointed appendages, and external skeletons of case-hardened chitin. Arthropods, in turn, are closely related to the nematodes, or roundworms—tiny but ubiquitous animals that include the parasitic agents of hookworm disease, elephantiasis, and trichinosis. These two phyla share features that were present in *their* last common ancestor, including unusual aspects of early development, molting cuticles, and nucleotide order in gene sequences. On the other hand, arthropods and nematodes could hardly look more different—the latter are little more than tiny cylinders tapered at both ends. Their last common ancestor must, therefore, have been a relatively simple organism, not, in detail, like a nematode and *certainly* unlike any living arthropod. The obvious differences between the last common ancestor of living species within a phylum and the last common ancestor of two sister phyla highlight a fundamental point about body plan evolution. The divergence of lineages depicted by branch points in phylogenies and the evolution of complex body plans *within* lineages constitute two separate phenomena. A host of biological changes took place between the divergence of the arthropod lineage and the last common ancestor of living arthropods.

The line of descent from the last common ancestor of arthropods and nematodes to animals with recognizably arthropod body plans is littered with extinct forms—perhaps forms with segmented bodies but not chitinous skeletons or jointed legs, or, later, forms with segments and chitin, but not jointed appendages. Biologists have a particular way of speaking about these evolutionary way stations. Two concepts are so important that they qualify as one more set of Jacob Marley facts.

The *crown* group of a phylum (or class, or any other clade) includes the last common ancestor of the phylum's living members and all its descendants (figure 11.3). Thus, at some moment in the latest Proterozoic or Early Cambrian, there lived a population of protoarthropods that split in two. Through time, one subpopulation gave rise to spiders, scorpions, and their relatives. The other evolved into crustaceans, insects, and *their* relatives. Some of the ur-population's descendants became extinct, themselves—for example, the scorpionlike eurypterids

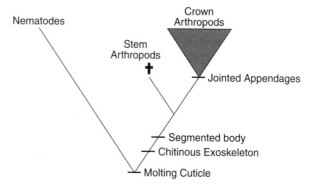

Figure 11.3. Diagram illustrating the concepts of stem and crown groups, as exemplified by arthropod evolution. See text for discussion.

that plied Paleozoic seaways. But with the origin and subsequent divergence of that founding population, the course of modern arthropod diversity was set.

Extinct way stations that bridge the gap between crown group arthropods and the divergence of the arthropod and nematode lines are called *stem* arthropods (figure 11.3). To comparative biologists, steeped in the details of living animals, stem groups and last common ancestors are hypothetical constructs inferred from phylogenetic trees. Not so for paleontologists. In our world of deep time and fossils, the ancestors of living animals are real organisms that swam, crawled, or just stood still in long-vanished oceans. We don't have to imagine them; we see them in rocks, preserved as skeletons and compressions if not actual flesh and blood. Paleontology, and paleontology alone, gives scientists access to the stem group species that illuminate the evolution of complex body plans. We *need* to understand stems and crowns to interpret the Cambrian record.

Ediacaran fossils such as *Kimberella* and *Spriggina*, which look both tantalizingly similar to and frustratingly different from modern animals, were probably stem bilaterians that lived in end-Proterozoic oceans. Not until the Cambrian, however, do we see crown group members of bilaterian phyla.

How can we understand the Cambrian emergence of a bilaterian world? The evolutionary possibilities inherent in the networks of genes that govern development—what Berkeley developmental biologist John Gerhart and Harvard cell biologist Marc Kirschner have called "evolvability"—contribute to an explanation, as do the amplifying effects of ecology and, possibly, environmental perturbation. These command our attention, but first we need to get a grip on the most fundamental aspect of Cambrian radiation—time.

Crown group members of bilaterian phyla didn't appear on January 1 of the Cambrian. As we saw in the Kotuikan cliffs, basal Cambrian rocks contain only a limited diversity of problematic remains, mostly small tubes made by wormlike animals and possible cnidarians. Stem groups that can be related to arthropods, or mollusks, or brachiopods appeared later, and their crown groups later yet. How much later has become clear over the past decade, as geochronologists, led by MIT's Sam Bowring, have dated volcanic ash beds layered among fossiliferous Cambrian rocks. Tracks and trails suggest that animals with jointed legs appeared within the first 10 million years of the Cambrian, but trilobites (plate 8) didn't make their entrance until the period was 20 million years old. And crown group crustaceans and chelicerates (the group that includes spiders and scorpions) may have no records older than about 511 million years, when the Cambrian had been under way for more than 30 million years.

Graham Budd and Sören Jensen, two leading paleontological Young Turks, have argued that the pattern just outlined for arthropod emergence fits other bilaterian phyla as well. For example, small caplike shells of calcium phosphate can be found in Siberian rocks formed when the Cambrian was about 13 million years old. These fossils are recognizable as brachiopods, but preserved details of shell structure show that their muscle systems and mantle tissues differed from those of any living lampshell (figure 11.4a). Probable crown members of the two great brachiopod lineages didn't enter the record for another 7 to 10 million years. Then there are the helicoplacoids, baglike fossils covered by spirally arranged plates of calcium carbonate, found in rocks as old as 520–525 million years (figure 11.4b). Feeding structures and the microscopic details of their skeletons clearly relate helicoplacoids to the phylum Echinodermata, but these strange fossils are quite unlike any echinoderm living today. Crown group echinoderms appear only later in the Cambrian, or even in the Ordovician, depending on how one interprets

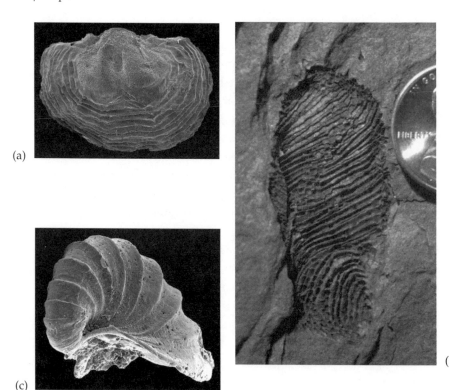

(a)

(b)

(c)

Figure 11.4. Cambrian fossils interpreted as stem groups of bilaterian phyla or classes. (a) an Early Cambrian brachiopod. (b) A helicoplacoid echinoderm. (c) A spirally coiled mollusk from basal Cambrian rocks. See text for discussion. (Image (a) courtesy of Leonid Popov; (b) courtesy of David Bottjer and Stephen Dornbos; (c) courtesy of Stefan Bengtson)

a handful of mid-Cambrian fossils. As a final example, consider the great phylum Mollusca. The Ediacaran fossil *Kimberella* may be an early way station in the evolution of mollusks, and minute spiral shells made by primitive mollusks are common in rocks formed a few million years after the beginning of the Cambrian (figure 11.4c). Tiny stem group clams and gastropods followed, about 15 million years into the period, but the shell beds of large clams, snails, and cephalopods familiar to all paleontologists didn't emerge as a conspicuous feature of sedimentary rocks until much later, in the Ordovician Period.

Paleontology's most famous fossils bring Cambrian evolution into sharp focus. The Burgess Shale is justly renowned for its remarkable

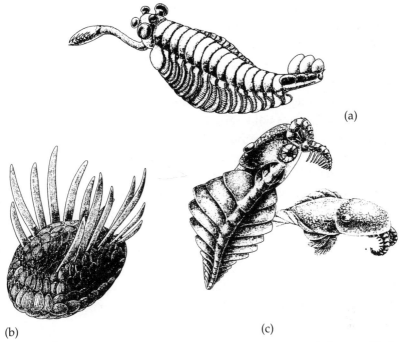

(a)

(b)   (c)

Figure 11.5. "Weird wonders" from the Middle Cambrian Burgess Shale. (a) *Opabinia*. (b) *Wiwaxia*. (c) *Anomalocaris*. (From S. J. Gould's *Wonderful Life*, W. W. Norton and Company, Inc., reproduced with permission)

store of Cambrian animals. Spectacularly detailed compressions of sponges, comb jellies, polychaete worms, priapulids, brachiopods, arthropods, and even a lancelet-like chordate tell us that by Burgess time, a great deal of body plan evolution had occurred within bilaterian phyla. But these remains share bedding surfaces with a host of mind-bending problematica that challenge paleontological understanding. The two-inch long *Opabinia* (figure 11.5a) had chitin-covered segments and feathery gills like any good arthropod, but it had no legs, and—worse—it sported five eyes and a grasping proboscis. The sluglike *Wiwaxia* (figure 11.5b), sallying forth in chain mail of chitinous scales, is equally bizarre. And so is *Anomalocaris* (figure 11.5c), a giant (up to two feet long!) predator distinguished by a segmented body with fanlike lobes instead of legs, but a pair of jointed appendages on the underside of its head to stuff food into its strange diaphragm-like mouth.

In *Wonderful Life*, Stephen Jay Gould focused intently on *Opabinia*, viewing it as key to the biological interpretation of Burgess fossils.[3] Steve paid particularly careful attention to those features that separate *Opabinia* from living animals—the odd proboscis and sci-fi eyes. As a result, he assigned this fossil to an extinct phylum, like Seilacher's vendobionts but younger (and within the fold of bilaterian animals). Steve likewise interpreted *Wiwaxia, Anomalocaris,* and other "weird wonders" as extinct body plans unlike anything seen in modern oceans.

Let us grant that some Burgess animals *are* very strange. But they aren't completely alien. After all, *Opabinia's* segmented body and chitinous external skeleton suggest evolutionary relationship to true arthropods. *Anomalocaris* had a segmented trunk, chitinous external skeleton, *and* jointed appendages, at least on its head. Weird or not, these fossils are stem groups that provide glimpses of how modern arthropods came to be—the concrete remains of forms mentioned earlier as hypothetical way stations in arthropod evolution.[4] And, as shown by Nick Butterfield, *Wiwaxia* scales have a microscopic structure that relates them to polychaete worms (and possibly, as well, to their evolutionary cousins, the mollusks). Even Burgess fossils once thought to represent crown groups of bilaterian phyla appear, on further investigation, to include stem taxa. For example, *Aysheaia pedunculata,* a small fossil long accepted as an early velvet worm, has a mouth and terminal appendages like those found today in a closely related phylum called the tardigrades, or water bears.

How old, then, are the stem-rich fossils of Burgess? No ash beds have been found in Burgess outcrops, but biostratigraphic correlation with other, well-dated successions indicates that these magnificent animals lived about 505 million years ago, nearly 40 million years after the start of the Cambrian Period.[5] The Burgess Shale, thus, stands as a monument

---

[3] Steve originally called his book *Homage to Opabinia,* but his editor, perhaps wisely, rejected this title in favor of *Wonderful Life,* an agreeable allusion to Jimmy Stewart and to all that is, well, wonderful about life.

[4] This point of view, first articulated by British paleontologists Derek Briggs, Richard Fortey, and Matthew Wills, receives extended treatment in Simon Conway Morris's book *Crucible of Creation: The Burgess Shale and the Rise of Animals.*

[5] Similar, and equally spectacular, fossils from Chenjiang, China, and Sirius Passet in northern Greenland are somewhat older. These deposits, which include the earliest fishlike animals, may have formed as early as 520 million years ago—still more than 20 million years after the Cambrian Period began.

to the persistence of stem group animals as much as to the emergence of crown group bilaterian phyla. Forty million years after the Cambrian began, evolutionary way stations still played a major role in the ecology of marine environments. For many bilaterian phyla, crown group species came to dominate marine communities only some 15 to 20 million years later, during dramatically renewed diversification in the Ordovician Period.

In summary, body plans recognizable as arthropods, brachiopods, mollusks, and even chordates took shape over a 10- to 30-million-year interval during the first half of the Cambrian (figure 11.6). Following that, continuing evolution through the remainder of the period fashioned the coordinated sets of features seen today in the crown groups of bilaterian phyla and classes. That's about 50 million years in all. Should we regard this time frame as uncomfortably short or uninterestingly long? Was there really a Cambrian Explosion?

Some have treated the issue as semantic—anything that plays out over tens of million of years cannot be "explosive," and if Cambrian animals didn't "explode," perhaps they did nothing at all out of the ordinary. Cambrian evolution was certainly not cartoonishly fast—no surprise there. But no one who has trekked through thick successions of Proterozoic shale or limestone can doubt that Cambrian events transformed the Earth. Cambrian body plan evolution may have taken 50 million years, but those 50 million years reshaped more than 3 *billion* years of biological history.

If we dismiss the notion that Cambrian evolution unfolded too slowly to be interesting, should we be worried that it all transpired too quickly? Do we need to posit some unique but poorly understood evolutionary process to explain the emergence of modern animals? I don't think so. The Cambrian Period contains plenty of time to accomplish what the Proterozoic didn't without invoking processes unknown to population geneticists—20 million years is a long time for organisms that produce a new generation every year or two. Explanations of Cambrian evolution must be sought elsewhere, at the place where development and ecology meet.

In tradition and approach, paleontology and molecular biology lie at opposite poles of the life sciences. Over the past decade, however, the continuum of biological inquiry has curved back on itself to form a circle,

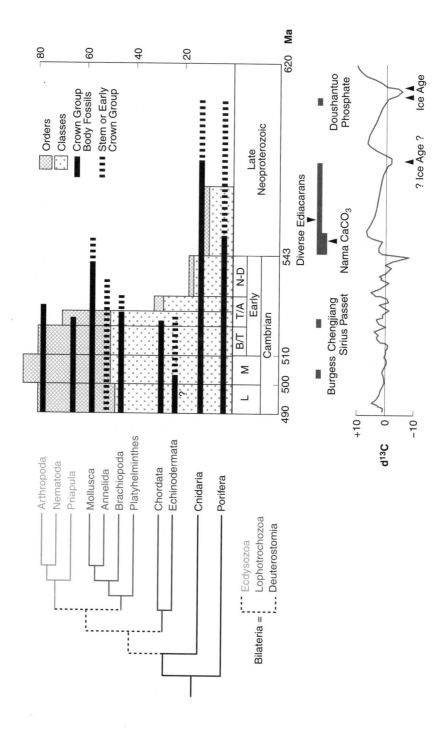

bringing paleontologists and molecular biologists into close and mutually informative contact. We've already seen evidence of this in the interplay between molecular phylogeny and deep Earth history. The link is further strengthened by the common interest of paleontologists and developmental biologists in body plan evolution. For, if fossils establish the stratigraphic pattern of early animal diversification, developmental genetics shows how this evolution could have been accomplished.

Let's return to the developmental tool kit, introduced earlier. Biologists have learned a great deal about development by studying that great workhorse of laboratory genetics, the fruit fly. As in all animals, fruit fly development proceeds by both the proliferation of cells and the specification of individual cell fates. The final action in cell differentiation is the expression of genes whose protein products alter the cytoskeleton and other cytoplasmic features to form functionally distinct cells such as neurons or muscle fibers. These proteins are indeed molecular carpenters, but which carpenters get tapped to finish any given cell is determined by "upstream" genes that regulate the overall developmental patterning of the fly.

The genetic cross talk that eventually determines individual cell identities begins in the earliest stage of development, before the fertilized egg even starts to divide. Proteins manufactured from RNA messages supplied by the egg's mother establish a concentration gradient from one end of the egg to the other. These proteins selectively promote or inhibit translation of other RNA messages, thereby defining the front and rear ends of the nascent body axis—a pattern that will be enhanced and elaborated as dividing cells begin to transcribe genes from their own nuclei. Continuing gene interactions specify ever smaller zones along the body of the developing embryo, eventually dividing it into a series of discrete segments—a hallmark of arthropod organization. Once genes have given fruit flies their segments, they start adding legs, wings,

---

Figure 11.6. A summary of animal phylogeny and Cambrian evolution, showing first-known appearances of stem and crown group members of extant phyla, as well as animal diversity through the Cambrian Period. To aid discussion in chapter 12, the record of carbon isotopes across the Proterozoic-Cambrian boundary is also shown. (Reprinted with permission from A. H. Knoll and S. B. Carroll, 1999. Early animal evolution: emerging views from comparative biology and geology. *Science* 284: 2129–2137. Copyright 1999 American Association for the Advancement of Science)

antennae, and eyes. Genes of the *Hox* family are expressed along the developing body axis in overlapping zones, and the developmental fate of each segment reflects the combination of *Hox* proteins at work in its cells. (The eight *Hox* genes of fruit flies cluster together on two segments of a single chromosome; remarkably, the linear arrangement of these genes corresponds to their spatial order of expression in the fly.)

Once segment fate has been specified, further genes initiate structural elaborations on individual segments. A gene called *Distal-less* initiates limb development, one pair on each segment. Constrained by *Hox* gene expression and guided by additional gene products, appendages on one head segment become antennae, while those on thoracic segments develop as legs. Limb formation is suppressed completely on the fly's abdomen. *Eyeless*, another well-studied gene, initiates eye development, while *tinman* signals the beginnings of the heart. (Genes deployed in development commonly have whimsical names. *Tinman* pays winking homage to Dorothy's metal-clad companion in Oz, but my favorite is *sonic hedgehog*, a gene deployed in the generation of limbs, teeth, hair follicles, and other features of vertebrate bodies.)

There is much more to fruit fly development than this—enough for an entire field to develop—but the observations presented here highlight a few key points. The patterning of the fly body begins in the egg and continues by means of gene interactions that specify ever more precise portions of the developing embryo. Mutations in the terminal genes that guide formation of specific cell types tend to have individually small effects, influencing eye color, bristle number, or the like. In contrast, mutations in the regulatory genes expressed earlier in developmental networks can have dramatic consequences. *Hox* gene mutations, for example, produce monstrous, and usually lethal, body forms: walking legs may sprout where antennae were meant to go, or wings may develop on two segments instead of one. Such mutants show how strongly *Hox* genes control the basic look of the fly.

All arthropods share a common set of *Hox* genes; nonetheless, they exhibit a dazzling diversity of form. In no small part, this diversity can be related to variations in the number, identity, and morphological particulars (for example, appendage type) of the segments that form the arthropod body. Beginning with research by Michalis Averof and Michael Akam in Cambridge, England, developmental biologists have shown that the expression patterns of *Hox* genes correlate well with these variations in segment form (figure 11.7). This correspondence suggests, not only that *Hox* and other regulatory genes guide arthropod development, but that *mutations* in these genes gave rise to the diversity of arthropod form seen today. Indeed, recent work in the laboratories of William McGinnis, at the University of California, San Diego, and Sean Carroll, at the University of Wisconsin, has shown how small changes in a *Hox* gene called *Ubx* governed the evolution of six-legged insects from their many-limbed crustacean ancestors. *Hox* genes in laboratory flies are beginning to put a molecular face on the Cambrian Explosion.

The story gets better. The mouse is another stalwart of laboratory research, and just as biologists have teased out the genetic circuitry of fruit fly development, they have come to understand a great deal about how mice develop. Comparison of fly and mouse genetics reveals a remarkable and unexpected correspondence. Not only do mice and flies both maintain a limited but versatile genetic tool kit for development; the two kits contain many of the same tools. *Hox* genes specify developmental fate from head to tail in mice, just as they do in flies—although a series of gene duplications have given vertebrates four sets of *Hox* genes, each equivalent to the single series found in arthropods. Genes closely related to *eyeless* and *Distal-less* induce eye and limb development in mice, as they do in flies. Even mouse hearts are initiated by a genetic homologue of *tinman*. The similarities can be astonishingly close: *eyeless*-like genes excised from a mouse can induce normal eye development in fruit flies. Of course, despite the similarity of their developmental tool kits, a mouse egg develops as a furry rodent while the fly's becomes a tiny dive-bomber. Similar genes but different shapes—the pattern found *within* the arthropods holds *among* phyla, as well.

Spurred by the genetic commonalities of flies, mice, and the nematode *Caenorhabdites elegans* (another experimental favorite), biologists have tracked the developmental tool kit across the animal kingdom. Sponges and cnidarians maintain relatively small complements of regulatory genes, but all bilaterians share the expanded tool kit first discovered in mice and fruit flies. This means that the essential genetic prerequisites for bilaterian diversification were present in the last common ancestor of all living bilaterian animals. Based on bilaterian traces in Ediacaran rocks from the White Sea region of Russia, this ancestor must have

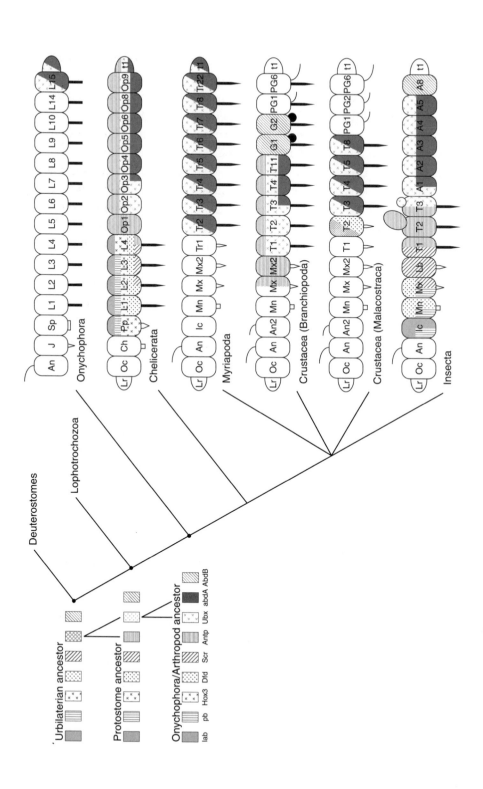

lived at least 555 million years ago. As the Cambrian began, then, the genetic engine of body plan evolution was already in place.

We can now begin to understand how Cambrian animals could have evolved more rapidly than Darwin envisioned. Mutations in regulatory genes made rapid diversification possible.

If mutations in regulatory genes fueled Cambrian diversification, can we further infer that such mutations were more common in the Cambrian than at other times? I doubt it. Genes mutate in present-day animal populations and do so at rates that are probably similar to those of the Cambrian. Most of these mutations are lethal—they produce animals that don't work. Some mutants survive in laboratory incubators, but we don't seem to see them in nature. Animals that function poorly simply can't compete in a world that is full of functionally sophisticated organisms.

This brings us to the crux of Cambrian evolution. To foment biological revolution, we need more than mutations; mutants must survive and reproduce, generating further variation on which natural selection can act. It is commonly assumed that evolutionary radiations begin with finely wrought innovations that take the world by storm. But this isn't the case—innovations become honed by natural selection only with the passage of time. Biological radiations *begin* when a permissive ecology allows poorly functioning novelties to persist.

By "permissive," I mean an ecological landscape in which competition for resources is rare or weak. (You don't have to be good to win the game of evolution; you only have to be better than the other players.) Permissive ecologies may arise because environmental change makes new physiologies possible, or because an evolutionary novelty allows organisms to exploit resources in a new way, however poorly. Environmental catastrophe provides still another route—populations that survive mass extinction may radiate in the ecological emptiness that follows, as mammals did after the dinosaurs disappeared. As

---

Figure 11.7. Expression of *Hox* genes along the body axis of a fruit fly (Insecta) and other arthropods, suggesting the molecular basis of morphological variation within the phylum. (Reprinted with permission from A. H. Knoll and S. B. Carroll, 1999. Early animal evolution: emerging views from comparative biology and geology. *Science* 284: 2129–2137. Copyright 1999 American Association for the Advancement of Science)

we'll see in chapter 12, latest Proterozoic and Cambrian history suggests that all three circumstances may have contributed to early animal diversification.

In 1996, Duke University biologist Greg Wray and his colleagues published a paper that set the paleontological world on its ear. Maybe fossils only preserve animals younger than 600 million years, they wrote, but the major bilaterian groups must have diverged from one another long before that, perhaps a billion years ago or even earlier. Wray's conclusion didn't come from geology, but rather from a particular reading of molecular biological data. "Molecular clock" estimates of evolutionary divergence times begin with the assumption that changes in the nucleotide sequences of genes accumulate in more-or-less clocklike fashion, with little variation through time or among taxa. If we accept this premise, we can measure the differences between genes in two species and use this to estimate when the lineages diverged from their last common ancestor. In fact, the starting assumption is contentious—many genes are known to violate it—but proponents of the molecular clock contend that violations can be recognized and eliminated from consideration.

Greg Wray's team compiled nucleotide sequences for specific genes drawn from many different vertebrate animal species. Then, they calculated the differences in gene sequence between many different pairs of species and plotted these on a graph against the times when the species parted genealogical company, as estimated from fossils. (Vertebrates were favored because bones leave a pretty good fossil record.) Surprisingly (at least to me), the data for some genes fell along a more-or-less straight line (figure 11.8), suggesting that nucleotide changes had indeed accumulated in clocklike fashion.

So far, we've only compared known genetic differences versus known times of evolutionary divergence. The contentious part comes next, with the assumption that the rates of molecular change calibrated for vertebrates can be extrapolated to deeper branches of the animal tree. Making this assumption, Wray's group measured gene sequence differences between pairs of protostome and deuterostome species and used these to estimate when the two great branches of bilaterian animals diverged from each other (figure 11.8). Genes for hemoglobin suggested a divergence time as early as 1.6 billion years, whereas a gene that codes for a

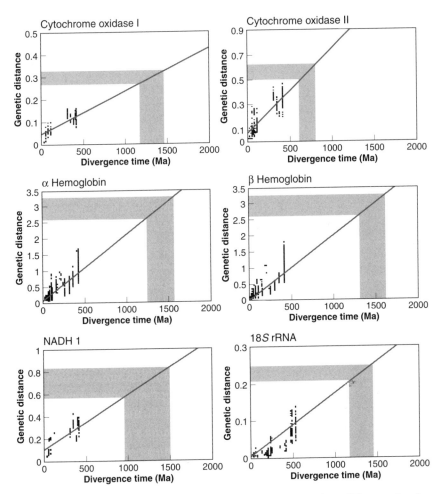

Figure 11.8. The molecular clock, illustrated using data from Wray and colleagues (1996). Shaded area shows measured genetic distances between deuterostome and ecdysozoan or lophotrochozoan species. Projection onto *x* axis estimates time since divergence from the species's last common ancestor. (Reprinted with permission from G. A. Wray, J. S. Levinton, and L. H. Shapiro, 1996. Molecular evidence for deep Precambrian divergences among metazoan phyla. *Science* 274: 568–573. Copyright 1996 American Association for the Advancement of Science)

cytochrome oxidase enzyme implied more recent divergence—perhaps 800 million years ago—but still long before the first Ediacaran fossils.

Stimulated by Wray's provocative results, other labs have taken up the challenge of molecular clocks and animal origins. Published estimates for bilaterian divergence range from 1.6 billion to 650 million years ago, reflecting differences in the genes and computations used in different studies.

Many biologists are uncomfortable with the idea that rates of molecular change can be extrapolated from vertebrates to animals as a whole. Given this concern and the truly wild variation in age estimates, it is tempting to conclude only that molecular clocks keep poor time. This may make paleontologists feel better, but it glosses over a potentially important point. Whatever their inconsistencies, *all* molecular clock estimates published to date indicate that animals began to diversify much earlier than fossils suggest.

Recently, Kevin Peterson and Carter Takacs of Dartmouth College have approached the molecular clock from another angle, basing their calibrations on echinoderm genes and fossils rather than vertebrates. They estimate that the last common ancestor of mice and fruit flies lived 540–600 million years ago, in close agreement with the fossil record. Many paleontologists like this estimate because it doesn't require an extended prehistory of the bilaterian supergroups (and suggests that we were right all along). But this doesn't let paleontologists off the hook. Perhaps the split between protostome and deuterostome bilaterians occurred only 540–600 million years ago, but according to the Tree of Life, bilaterians and cnidarians must have parted company earlier—and the common ancestors of bilaterians and cnidarians must have split from sponges earlier still. Peterson and Takacs estimate that the early branches of the animal tree formed 700–750 million years ago. The oldest known sponge and cnidarians, however, are less than 600 million years old. Thus, even the conservative molecular clock estimates of Peterson and Takacs require up to 150 million years of animal history not yet recognized in rocks.

How do we reconcile molecular clocks that call for early animal divergence with a fossil record that identifies metazoans as evolutionary latecomers? Andrew Smith, an accomplished switch-hitter in molecular phylogeny and paleontology, has outlined the options for reconciliation. There are only three.

Perhaps gene sequences diverged at unusually high rates early in animal evolution, undermining efforts to read time from molecules. The idea that genes might evolve rapidly during early diversification isn't absurd or even ad hoc—evidence exists for rapid gene evolution during the explosive radiation of some younger groups. Thus, if gene sequences in Phanerozoic vertebrates diverged by, say, 1 percent every 50 million years, while genes in the earliest animals diverged by 1 percent in 10 million years, extrapolation of vertebrate rates would seriously overestimate the divergence times for early branches on the animal tree. That said, in cases where genes are known to have evolved rapidly early on and then slowed down, their change of pace has left a recognizable signal in the relative abundances of the four nucleotides that spell out gene messages. Smith scrutinized available data on animal gene divergence and found no evidence of evolutionary rate change.

Perhaps, Smith continued, genes are telling us the truth and paleontology has misled us. That is, maybe we haven't sampled the geological record carefully enough and so missed animal fossils that sit in late Proterozoic rocks. Smith is modestly enthusiastic about this possibility, but having spent the better part of twenty years tromping around late Proterozoic rocks, I'm skeptical of the idea that by looking harder we'll find Cambrian-like fossils in older beds. Trace fossils illustrate the point. It isn't just that animal tracks and trails first appear in latest Proterozoic rocks, it is that they appear *everywhere* once they enter the record. In older rocks, on the other hand, you can search for a very long time without seeing anything.[6] Exceptional windows on the past like those of Doushantuo and Nama reinforce the idea that whatever we've over-

---

[6] Claims of older animal fossils surface every few years. The most celebrated candidates are sinuous impressions reported by Dolf Seilacher in rocks now known to be 1.6 billion years old. Dolf's structures may be biological, but features such as branching pattern suggest to me that they are more likely to be algal than animal. More to the point, I see no evidence to connect these structures stratigraphically to the record of abundant trace fossils that begins 555 million years ago or phylogenetically to the animals that made those younger trackways.

looked in late Proterozoic rocks, it isn't large, complicated animals like those of the Cambrian.

As Smith points out, the third alternative is that molecular biology and paleontology are both right, but are telling us different things. This takes us back to the point made earlier that body plan evolution within a group is different from (and postdates) the genealogical divergence of a group from its relatives. Accurate or not, molecular clocks estimate times of evolutionary branching. Fossils document body plan evolution within animal phyla.

Given the phylogenetic inference that major eukaryotic groups diverged rapidly and fossil evidence that multicellular red algae existed more than a billion years ago, it isn't crazy to conjecture that stem group animals emerged early as well. One simply has to accept that the earliest stem cnidarians and bilaterians were rare, gossamer, or minute organisms not likely to be preserved (or at least recognized) in the fossil record. But shared features of genetics and morphology place some limits on our speculation. Early cnidarians might have resembled modern *Hydra*, a tiny (up to one centimeter tall and rarely more than a millimeter wide) polyp that secretes no skeleton and leaves no mark on the sediment surface. Such animals would rarely if ever show up as fossils. In contrast, the last common ancestor of protostomes and deuterostomes was large and complicated enough to have left its calling card in sediments, at least if it moved across the seafloor or formed a preservable organic cuticle. Of course, those are big ifs—nematodes, for example, have probably been abundant throughout the Phanerozoic Eon, but they have left hardly any recognizable fossils (figure 11.9).

Accepting the qualitative if not necessarily the quantitative claims of molecular clock hypotheses also requires that we view Ediacaran events in a specific way—as the independent evolution of large size within major branches of animals that separated earlier. Why, we must ask, did large animals with their complicated body plans take shape so long after their tiny ancestors began to diversify?

At present, the problem remains unresolved. Paleontological tests of molecular clock hypotheses will require that we scour late Proterozoic sedimentary rocks in search of tiny but distinctive phosphatized fossils like those in Doushantuo rocks. Many are looking; as yet, no one has found anything convincing. But we also need to think hard about envi-

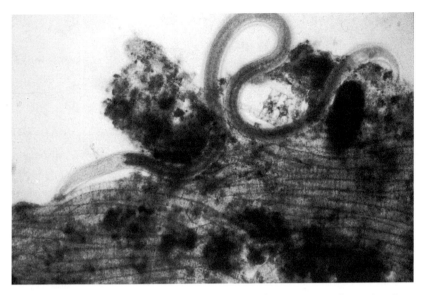

Figure 11.9. This sinuous tube, tapered at both ends, is a nematode—a tiny (less than a millimeter long) animal found almost ubiquitously in present-day environments. Nematodes have complex tissues and organs, but almost never fossilize. For comparison, the filaments in the lower part of the figure are sulfur-oxidizing bacteria! (Photo courtesy of Andreas Teske)

ronmental events that might have stimulated animal evolution, or more accurately, the evolution of large, preservable animals, 600–580 million years ago. Moreover, recalling fossils introduced in earlier chapters, we must explain the *simultaneous* emergence of diverse seaweeds and planktonic algae, large protozoans *and* large animals at this time. We need to look carefully at the momentous physical upheavals that shook the late Proterozoic world.

12 Dynamic Earth, Permissive Ecology

Supercontinental breakup, globe-swaddling ice, rising oxygen levels, and short-lived environmental perturbation at the Proterozoic-Cambrian boundary—great events in Earth's planetary history framed the early evolution of animals, generating successive waves of permissive ecology that fueled metazoan diversification.

For more than a century, paleontologists have scrambled across rock faces in search of early animals. Until recently, however, we've labored alone—other scientists had their own problems and their own agendas. Now, as already seen, paleontology has found an ally in molecular biology. And another partner, equally important, has emerged from our traditional home base of geology. Increasingly, Earth scientists are working to understand how rocks, life, air, and water interact to produce the environment around us. Much of this research, christened Earth system science, is motivated by concerns about our environmental future. But geochemists and climatologists have also begun to study Earth's environmental *past*, allowing us, for the first time, to link early animal evolution to late Proterozoic and Cambrian environmental history. And what a history it is turning out to be.

In Spitsbergen, the fossil-packed cherts of the Akademikerbreen Group are separated from younger Cambrian rocks by thick beds of tillite, the coarse and poorly sorted sediment deposited by glaciers. As observed in chapter 9, tillites also occur just below the extraordinary Doushantuo fossils of southern China. And in Australia, glacial rocks lie just beneath the thick sedimentary pile that contains Ediacaran fossils near its top. The same stratigraphic pattern can be found in subhimalayan India; in Eu-

Figure 12.1. Coarse and poorly sorted sedimentary rocks of the Numees Tillite, Namibia. Similar rocks seen around the world document widespread glaciation on the late Proterozoic Earth.

ropean Russia; in Norway, Namibia, and Newfoundland; in the Rocky Mountains, from Death Valley to northern Canada; even in Boston Harbor (figure 12.1). Ice heralded the age of animals.

Brian Harland—the same Brian Harland who invited me to work in Spitsbergen—was the first to recognize the implications of these widespread tillites, proposing in 1964 that Earth experienced a global ice age near the end of the Proterozoic Eon. We tend to be impressed (and rightly so) by the fact that 18,000 years ago Boston and Chicago lay beneath glacial ice, but Pleistocene ice sheets never extended south of Long Island, even at the height of glaciation, leaving much of North America ice-free. In contrast, if Harland was correct, *most* landmasses on the late Proterozoic Earth must have been covered by ice.

Decades of careful research have confirmed Harland's proposal. In fact, continental glaciers spread across the globe more than once, although the exact number of late Proterozoic ice ages remains contentious. Noting that only two tillite successions are found in most regions, some stratigraphers believe that the Earth froze twice. On the other hand, applied to Phanerozoic rocks, the same approach would suggest that our planet endured two ice ages in the past 500 million years, when in fact there were three.[1] Only precise radiometric dates can solve the problem. My own view is that Earth chilled at least four times during the late Proterozoic: an initial event a bit earlier than 765 million years ago and arguably confined to Africa, two truly global ice ages 710 ± 20 and a bit more than 600 million years ago, and at least one last (relatively small) hurrah before the Cambrian began.

In general, sedimentary geologists associate glacial rocks with cold climate and carbonate accumulation with warmth, but, as seen in Spitsbergen, late Proterozoic tillites commonly lie sandwiched between carbonate-rich successions. In particular, late Proterozoic glacial rocks are nearly always capped by distinctive carbonate beds that display unusual sedimentary features, including sheaves of slender crystals like those found in Archean and earliest Proterozoic limestones (figure 12.2). In many places, you can place a knife blade at the sharp contact between glacial rocks and the blanketing cap carbonates.

Our research in Spitsbergen turned up another unusual feature of these cap carbonates, in this case a chemical oddity. Recall from chapters 3 and 6 that two different types of information can be gleaned from carbon isotopes in sedimentary rocks. *Differences* in the isotopic compositions of limestone and organic matter reflect the metabolisms of organisms in the local ecosystem. On the other hand, the *absolute* ratio of ^{13}C to ^{12}C in limestones and dolomites allows us to estimate the relative contributions of carbonate and organic matter to sedimentary carbon burial at the time the rocks formed—higher C-isotopic values imply higher rates of organic carbon burial (figure 6.6). Thick carbonate successions below late Proterozoic glacial rocks generally have unusually high C-isotopic ratios, nearly matching the extremes found in 2.4–2.2-billion-year-old rocks and exceeding anything else seen in the geologic record.

---

[1] A relatively brief ice age centered on northern Africa took place near the end of the Ordovician Period, some 440 million years ago. Continental ice sheets returned late in the Paleozoic Era, covering much of Gondwana from about 355 to 280 million years ago. Glaciers began to expand for a third time 33 million years ago, in Antarctica, although ice sheets spread across northern continents only within the last 2 million years. No one region preserves a sedimentary record of all three events.

Figure 12.2. A cap carbonate above the Numees Tillite, Namibia. Cap carbonates characteristically show unusual bedding features, including the thin and contorted laminations seen here and fans of crystals deposited directly on the seafloor.

In contrast, the C-isotopic compositions of cap carbonates fall to extremely low values. This chemical pattern occurs not only in Spitsbergen but worldwide (with an additional twist, discussed below). Moreover, it applies to each major ice age.

Late Proterozoic ice ages, thus, can be linked to unusual behavior of the global carbon cycle. How do we explain these chemical fluctuations, and what do they tell us about the late Proterozoic world?

Geology helps to explain our isotopic data. The onset of unusually high C-isotopic ratios coincides with the rifting and breakup of one or more late Proterozoic supercontinents. As great continents broke apart, narrow seas opened, perhaps facilitating the burial of organic matter in rapidly accumulating sediments. That is to say, *tectonic* changes may explain the *chemical* observation of high C-isotopic values. High rates of organic carbon burial may, in turn, have helped to keep atmospheric CO_2 at relatively low levels, cooling global climate and leaving the Earth vulnerable to glaciation. The intricate linkage of Earth and environment implied by these relationships is what Earth system science is all about.

The low C-isotopic values in cap carbonates have been interpreted in several ways. One possibility is that algae and cyanobacteria were scarce in the postglacial ocean and the burial rate of organic matter correspondingly low. Another is that cap carbonates accumulated at extremely rapid rates, swamping any influence of organic carbon burial on the C-isotopic record. We might also conjecture that as glacial ice retreated, methane (which has very low C-isotopic values) was belched from the margins of warming continents.

There's more. Conspicuous carbonate precipitates were not the only archaic sedimentary features reprised in association with late Proterozoic glaciation. Iron formations returned, as well. As documented by Grant Young, a geologist at the University of Western Ontario, iron formations occur among late Proterozoic tillites throughout the world, especially those formed during the great ice age circa 710 million years ago. We explained earlier iron formations as the products of dissolved iron transport in deep oceans that lacked oxygen or sulfide. How could such oceans return more than a billion years after they disappeared, seemingly for good?

We will return to the question of ice and iron, but there is one more aspect of late Proterozoic glaciation that requires attention, because it may provide the key to everything else: at least some tillites formed at sea level near the late Proterozoic equator. This was not your Cro-Magnon ancestors' ice age.

How do we know this? Given that continents have migrated through time, rafted by tectonic plates, how can we tell that rocks formed deep in our planet's history accumulated in the tropics? The answer lies in the magnetic properties of sedimentary and volcanic rocks. When rocks form, iron-bearing minerals crystallize in alignment with the Earth's magnetic field. Frozen in place, this magnetic orientation can be preserved through geologic time, enabling geologists to determine the latitude—but not longitude—at which ancient rocks originated. (Magnetic orientation can also be reset by later events, so geologists must be exceedingly careful in interpreting paleomagnetic data.) Meticulous studies of glacial rocks in southern Australia and western North America show that these beds formed within 10° of the late Proterozoic equator. At circa 40° paleolatitude, the Nantuo tillite in China is among the most *poleward* of the admittedly few tillites whose magnetic signatures have been reliably measured.

In 1992, Joe Kirschvink, a gifted maverick in Caltech's geology department, painted an extraordinary picture of late Proterozoic ice ages. According to Joe, glaciation may have begun conventionally, at high latitudes or elevations, but as ice sheets expanded toward the equator, Earth's climate system approached and then crossed a critical threshold. Ice reflects sunlight back into space (in the language of climatology, it has a high albedo), cooling the planet as it expands. Cold, in turn, facilitates further glacial growth. Thus, there is a positive feedback associated with waxing ice sheets. In Kirschvink's view, once glaciers expanded to within about 30° of the equator, runaway refrigeration ensued, covering the Earth with ice in as little as a few thousand years. Ice sheets stretched from pole to pole, and sea ice blanketed the oceans, forming, in Kirschvink's evocative phrase, a Snowball Earth.

At first, few people took Kirschvink's proposal seriously. For one thing, it required that we set aside accepted models based on Pleistocene climate in favor of a radical alternative, a move rarely favored by innately conservative scientists. Moreover, the Snowball Earth presents a fundamental problem—once the planet gets into this state, it's hard to get back out.

In 1998, however, stock in the Snowball Earth appreciated dramatically. During a series of late-night conversations in the Harvard Earth Science building, geologist Paul Hoffman and geochemist Dan Schrag conceived of a way to reconcile geological observations of late Proterozoic glaciation with Kirschvink's vision of a frozen planet. In particular, they interpreted the geology and chemistry of cap carbonates as evidence for Joe's proposed escape route from the Snowball's icy grip.

In Hoffman and Schrag's reworking of the Snowball Earth hypothesis, late Proterozoic ice ages began, much as Kirschvink suggested, with conventional glaciers that crossed a critical latitudinal threshold and then rapidly covered the planet. The global lid of ice nearly shut down primary production, accounting for a putatively extended interval of low C-isotopic values—which Hoffman and colleagues found begins just *below* the tillites at least for one ice age. The icy veneer further prevented oxygen from diffusing into seawater from the atmosphere, resulting in anoxic deep oceans. In the absence of primary production, rates of sulfate reduction also slowed, limiting the production of H_2S and, therefore—for the first time since 1.8 billion years ago—allowing iron to build up in oceanic deep waters.

Cold and ice also shut down continental weathering. Thus, glacial ice nearly stopped the two principal processes that remove carbon dioxide from air. But ice couldn't stop the main engine by which CO_2 is *added* to the atmosphere—volcanism. As a result, CO_2 levels gradually rose in the air above the icy wastes. Carbon dioxide is an important greenhouse gas, but Hoffman and Schrag estimate that to vanquish the global ice sheets, atmospheric CO_2 must have risen to levels as much as 300 to 400 times higher than today's. It takes time to accumulate such massive stores of carbon dioxide—Hoffman and Schrag suggest several million years, in line with their estimates of glacial duration—but once a critical threshold (again) was passed, deglaciation would have been nearly instantaneous, melting the vast ice sheets (causing a sharp rise in sea level) and catapulting ocean temperature to 40°C or higher, much hotter than the warmest seas today. Chemical weathering on the hot postglacial Earth would have proceeded at record high pace, feeding large amounts of calcium into the oceans, where cap carbonates precipitated. Weathering also would have drawn CO_2 from the atmosphere, hastening our planet's return to normalcy.

What about Snowball biology? If the apocalyptic scenario of Hoffman and Schrag is to be believed, previously productive habitats would have disappeared as ice expanded, eventually restricting most marine organisms to small refuges around emergent hydrothermal vents like present-day Iceland. Then, things would have gotten really bad. Having narrowly escaped an end by ice, the biological world would have been scorched by "fire" as the oceans heated up to temperatures that few eukaryotes can tolerate for long intervals. Despite this, Hoffman and Schrag suggest that the Snowball Earth and its aftermath provided the crucible that forged animal life. This belief rests largely on the stratigraphic appearance of Ediacaran animals above late Proterozoic tillites, but it is bolstered by the suggestion, much debated in genetic circles, that extreme environmental stress can induce mutations that fuel biological innovation.

The Snowball Earth qualifies as a Big Idea, taking in a remarkable sweep of climatic history, geochemistry, and biology. Not surprisingly, then, it has spawned vigorous argument, as Big Ideas usually do. Not all aspects of the debate are germane to this discussion, but as paleontologists we do need to ask two questions: did some form of Snowball really happen, and if it did, what might we reasonably conclude about its evolutionary consequences?

Two issues illustrate (but do not completely encompass) the nature of current disputes. First, there are questions about water during late Proterozoic ice ages. In the original Snowball scenario, sea ice as much as half a mile thick severely restricts the transfer of water vapor from ocean to atmosphere, slowing Earth's hydrological cycle almost to a standstill. Yet, those Australian tillites that formed near the late Proterozoic equator reach a thickness of more than three thousand feet, telling us that ice sheets must have continued to grow for a long time after glaciation reached the tropics. Given that low-latitude sea ice both begins and ends rapidly in the Snowball scenario, it isn't easy to reconcile these tillites with hydrological shutdown. Now, we can relent on sea ice thickness and envision oceans with no more than a thin (maybe a few feet) veneer of ice that would crack and break, allowing the hydrological cycle to continue. But, if we accept this, we must allow carbon dioxide exchange between air and sea, potentially limiting the hypothesized buildup of atmospheric CO_2.

A second issue concerns the carbon isotopic record outlined above. C-isotopic values may be low both before and after ice ages, but in the evolving Snowball scenario, they are low for different reasons. Along with graduate student Pippa Halverson and Yale University geochemist Robert Berner, Schrag and Hoffman hypothesize that methane leakage from organic-rich sediments controlled immediately preglacial climates. Biogenic methane is strongly enriched in ^{12}C (recall chapter 6), and molecule for molecule, it is far more effective as a greenhouse gas than carbon dioxide. Schrag, Hoffman, and colleagues suggest that, in essence, the Earth became addicted to methane for keeping warm, and when for some reason the supply was shut off, temperatures plunged and ice expanded. (Note that this proposal can work only if oxygen levels remained low on the late Proterozoic Earth; today, methane released slowly from seafloor sediments would react with O_2 as it ascended through the water column, delivering carbon dioxide to the atmosphere. We'll return to the question of oxygen below.) In contrast, Schrag and Hoffman ascribe postglacial C-isotopic values to superrapid carbonate precipitation.

Because these hypotheses explain preglacial and postglacial C-isotopic values in different ways, they make no prediction about carbon chemistry *during* the ice ages. Many workers have assumed that C-isotopic values were continuously low through the glacial epoch—hence the hypothesis that primary production fell to extremely low levels—but recent analyses of relatively rare carbonate beds *within* tillite sections call this assumption into question. Three leading Snowball skeptics, Martin Kennedy of the University of California at Riverside, Tony Prave of Aberdeen University in Scotland, and Columbia University's Nick Christie-Blick (who once gave a lecture at MIT titled "A Neoproterozoic Snowjob," leaving no ambiguity about his views), have systematically sampled intraglacial carbonate beds from several continents. These rocks turn out to have C-isotopic values that are rather normal for limestones and dolomites of any age. It might be argued that intraglacial carbonates formed by the redeposition of older, preglacial rocks, but the beds include oolites that must have precipitated from ice age seawater. Remembering that the C-isotopic composition of carbonates reflects *proportional* rates of organic- and carbonate-carbon burial, we can conclude either that carbonate deposition plummeted along with primary production during global glaciation or that primary production didn't drop off so much after all.[2]

In fact, most aspects of the late Proterozoic glacial record can be interpreted in more than one way, including the proposed extent of ice cover. Tom Crowley, now at Duke University, has investigated late Proterozoic glaciation using climate models of the type developed to explore twenty-first-century global warming. In his models, ice sheets spread rapidly once they reach a paleolatitude of 30–40°, and sea ice ex-

---

[2] These data potentially undermine another argument advanced for the Snowball scenario. Alternative explanations for low C-isotopic values in cap carbonates include a postglacial methane burst and the upwelling of deep waters enriched in ^{12}C—mechanisms that can influence the composition of seawater for a few hundred thousand years, at most. If the C-isotopic value of seawater remained low for several million years, from the onset of global glaciation until its aftermath, hypotheses that call for upwelling or methane release must be rejected. If, on the other hand, low isotopic values began only with deglaciation and ended shortly thereafter, carbon chemistry does not necessarily favor any one explanation over the others.

pands to cover much of the world's oceans—score one for the Snowball Earth. But unlike the full-tilt Snowball scenario, some iterations of Crowley's model leave extensive areas of equatorial ocean ice-free. Another difference: Crowley's ice begins to retreat when atmospheric CO_2 reaches a modest four to five times preglacial levels.

Who is right? Nobody knows, in part because we still lack observations or measurements that might help us to choose among competing plausibilities. I like many features of the Snowball hypothesis. Nonetheless, I confess a preference for milder, "slushball" variants of late Proterozoic climatic history, versions that begin and end less catastrophically and leave a bit of open water in the oceans, because I find them easier to reconcile with what I see in the field and because they don't require ad hoc assumptions to explain the survival of diverse eukaryotic organisms. (Bacteria and Archaea are less informative because, as outlined in chapter 7, prokaryotic taxa are hard to eliminate under almost any circumstances.)

The final chapters in Snowball research have yet to be written. But we know enough to be clear on what is probably the key point in all of this: even in the mildest permissible scenarios, ice covers much of our planet's oceans and most of its continental shelves. No resolution of the current debate will restore the comfortable idea that late Proterozoic ice ages were like the Pleistocene, only older. Late Proterozoic glaciation was extraordinary, and it must have left its mark on contemporary biology.

In chapter 9, we discussed the early fossil record of eukaryotic organisms. Red algae appeared well before the onset of late Proterozoic ice ages, and so must have survived the vicissitudes of both glaciation and deglaciation. Green algae did, as well, along with relatives of brown algae, dinoflagellates, ciliates, and testate amoebas. If molecular clocks have any merit at all, even microsocopic animals must have weathered some or all of these climatic storms.

The list of survivors expands further when we add inferences from the Tree of Life. For example, fossils of testate amoebae in 750-million-year-old rocks require, at a minimum, that the common ancestor of fungi and animals was also present, because the evolution of identifiable characters in testate amoebae must logically have come after the phylogenetic split between this group and the animals + fungi. In fact, *most* major groups of present-day eukaryotes must have been present before the late Proterozoic ice ages began—a lot of lineages survived climatic upheaval.

This perspective on ice age biology indicates that refuges must have been numerous, widespread, and persistent during the worst of the glaciers—explaining my preference for relatively mild paleoclimatic scenarios. On the other hand, tabulating ice age survivors doesn't fully address questions about ice age extinctions. After all, while the phylum Brachiopoda survived the great Permo-Triassic mass extinction 251 million years ago, more than 90 percent of all brachiopod *species* disappeared. Many microscopic eukaryotes leave no identifiable fossils, so our ability to evaluate the magnitude of ice-related extinctions is limited. Nonetheless, by tracking the records of eukaryotes that do fossilize, Gonzalo Vidal and I observed years ago that many protists failed to survive late Proterozoic glaciation—climatic shifts *did* prune the eukaryotic tree. That said, the most conspicuous plankton extinction in the Proterozoic record occurred just after the Doushantuo Formation was deposited (chapter 9), perhaps associated with one last cold snap, but *not* in conjunction with global glaciation. Snowball ice was not the only influence on late Proterozoic life.

If patterns of extinction and survival allow a range of ice age scenarios, what should we make of attempts to link late Proterozoic glaciers with biological innovation?

Research on stress-induced mutations remains in intriguing infancy, but as yet we have little evidence that stress facilitates mutations that lie beyond the realm of more mundane genetic processes or that these mutations adapt animals for conditions beyond those that induced the stress. Of course, in "slushball" scenarios for late Proterozoic climate change, surviving populations need not have been subjected to extremely harsh glacial or postglacial conditions. Moreover, the numbers of survivors needn't have been small. When organisms are tiny, many individuals can fit into a small area—a square meter of beach sand, for example, can harbor millions of nematodes.

The principal influence of global glaciation was probably ecological, regardless of which ice age scenario we choose to favor. Recall the argument, introduced last chapter, that permissive ecology foments biological revolution. The growth of globe-swaddling ice sheets would have removed biology from much of the Earth's surface, and when, in due course, the ice receded, huge areas of real estate would have become available for recolonization. Successful colonists likely experienced little competition, permitting novel and, perhaps, poorly functioning variants to survive. Genetic variation is necessary for evolutionary radiation, but is not sufficient. Populations need Lebensraum where nascent novelties can survive and reproduce. That is just what the decay of late Proterozoic ice ages provided.

However, there is a problem. At the beginning of this chapter, I stated that the late Proterozoic Earth experienced at least four ice ages, two of global dimensions. But Ediacaran animals and diverse algae follow only the last ice age—earlier Proterozoic glaciation did not engender recognizable evolutionary innovations. Ice ages were not magic wands of evolution.

We have to ask what was different about the aftermaths of younger and older ice ages. I believe that oxygen made all the difference.

In 1959, J. R. Nursall, a zoologist at the University of Alberta, proposed that animals appeared so late in the evolutionary day because only in the latest Proterozoic did our atmosphere accumulate enough oxygen to sustain metazoan physiology. Nursall based his hypothesis strictly on comparative biology—I'm not sure he ever knowingly looked at a Proterozoic rock—nonetheless, his proposal has enjoyed episodic popularity among paleontologists. Only in the past decade, however, has Nursall's idea garnered empirical support from geology.

There is no question about animals' need for oxygen. In modern marine basins, the abundance and diversity of animal species declines precipitately as we approach anoxic waters. Small animals persist better than large—nematodes and other tiny animals that inhabit the spaces between sand grains on the seafloor have especially low oxygen needs. Heavily skeletonized species fare most poorly; few armored animals live in oxygen-starved environments.

As early as 1919, August Krogh showed that when marine animals depend on diffusion to supply their tissues with oxygen, body size is limited by the amount of oxygen in surrounding waters. Krogh's biophysical rule can be circumvented in many ways—animals like cnidarians (and, probably, the vendobiontids in Ediacaran assemblages) achieve large size by draping thin layers of metabolically active tissue around inert "jelly" or fluid. Circulation of body fluids and specialized respiratory organs (gills or lungs) help as well, carrying oxygen efficiently to distant tissues. Nonetheless, on the Proterozoic Earth, before animals evolved sophisticated circulatory systems, oxygen levels must have determined the effective sizes of animals.[3]

The application to late Proterozoic biological history is obvious. Microscopic animals with scant oxygen requirements could have plied Proterozoic seas long before the Ediacaran epoch. Only with a latest Proterozoic rise in oxygen, however, did macroscopic (and, hence, easily fossilizable) animals become possible.

The oxygen hypothesis reconciles paleontology and molecular clocks, and it explains the simultaneous radiations of animals and algae in latest Proterozoic oceans—recall from chapter 9 that only in oxygen-rich oceans would large-celled eukaryotic algae have gained prominence across continental shelves. As outlined above, a latest Proterozoic rise in oxygen also makes Schrag and Hoffman's proposal for methane-driven glacial initiation possible and, if that model is correct, explains why global ice ages never returned in the age of animals.

Did any such event leave its mark in the geochemical record? The answer is yes. When John Hayes, Jay Kaufman, another lab alumnus, and I first documented high C-isotopic values in late Proterozoic carbonates, we recognized this as a smoking gun for oxygen influx. As discussed in chapter 6, photosynthesizing cyanobacteria and algae use carbon dioxide and water to produce both organic molecules and oxygen. Respiring organisms react the organic matter and oxygen back to carbon dioxide and water. So, as long as photosynthesis and respiration remain in balance, the environment doesn't change. Burial of organic matter breaks this metabolic couple, making it possible for oxygen to accumulate in the atmosphere and oceans. High C-isotopic values imply that in late Proterozoic oceans, organic matter was buried at unusually high rates. Indeed, when Lou Derry, Jay Kaufman, and Stein Jacobsen used

---

[3] Oxygen availability places biochemical as well as biophysical constraints on animals. For example, Kenneth Towe of the Smithsonian Institution noted years ago that the biosynthesis of collagen, the extracellular matrix protein par excellence, requires O_2 in relatively high concentrations.

C-isotopic and other geochemical data to model late Proterozoic atmospheric change, they concluded that oxygen increased strongly just after the younger of the global ice ages—supplying an answer to the question of why life responded differently in the aftermaths of earlier and later glaciations on the late Proterozoic Earth.

Independent evidence for latest Proterozoic oxygen increase comes from the sulfur isotopic record. In chapter 6, I outlined the thinking that led Don Canfield to propose that mid-Proterozoic oceans featured moderate oxygen at the surface and hydrogen sulfide at depth. One of Don's key observations was that S-isotopic fractionation didn't reach modern values until the end of the eon. The sulfidic ocean returns here—now, however, seen in its closing moments. Only as the Proterozoic Eon ended did sulfur-bearing minerals begin to record the fractionation expected in oceans full of oxygen. Sulfur isotopes also suggest that sulfate concentrations rose to near modern levels at this time, consistent with an overall increase in Earth's surface oxidation level.

Thus, geochemistry increasingly supports Nursall's hunch that animal evolution was stirred by oxygen. And with more oxygen, a new world began to emerge. Seaweeds and planktonic algae diversified across continental shelves. Among animals, developmental mutations favoring large body size stopped being lethal and began to be advantageous, introducing new functional possibilities. By 555 million years ago, large size had evolved in colonial protozoans, sponges, cnidarians (and vendobionts), and stem bilaterians.

Perhaps there is nothing more to explain. Perhaps, with the genetic tool kit in place, removal of the oxygen barrier simply allowed animal life to unfold. But there is one more event to explore.

Namibian fossils demonstrate that Ediacaran organisms reigned over marine communities until the very end of the Proterozoic Eon. Conversely, the Kotuikan cliffs show that bilaterian diversity (literally) took shape later, as the Cambrian began. Intriguingly, at the boundary between these two faunas, C-isotopic values fall through the floor, signaling a large, but short-lived, perturbation in Earth's carbon cycle.

Around the world, carbonate rocks near the Proterozoic-Cambrian boundary display C-isotopic values as low as or lower than those in the cap carbonates that cover late Proterozoic glacial beds (figure 11.6).

However, there is little evidence of tillites at the boundary, nor do we find any of the unusual sedimentary features that mark postglacial caps. Radiometric dates tell us that the isotopic anomaly came and went in less than a million years. And research on trace elements (especially uranium enrichment) by Japanese geochemists Hiroto Kimura and Yoshio Watanabe suggests that coastal oceans were transiently starved of oxygen.

What could have caused such a perturbation? We don't know for sure, but the candidate list is short, and all plausible explanations imply bad times for biology. Moreover, similar chemistry marks a major event in more recent Earth history: the great mass extinction at the boundary between the Permian and Triassic periods. Permo-Triassic catastrophe has been blamed on great bursts of methane from icy stores on continental slopes or catastrophic overturn of the hemisphere-scale Panthalassic Ocean, bringing oxygen-poor but CO_2-rich waters to the surface. Impact by a meteor or comet also attracts interest, although evidence for extraterrestrial influence on end-Permian extinction remains controversial. The same suspects form the lineup for end-Proterozoic environmental perturbation, but, as in the Permian, we don't yet know which party, if any, is guilty.

Could mass extinction explain the stratigraphic break and morphological gulf between Ediacaran and Cambrian animals? I think that it can.[4] In fact, marine life across the Proterozoic-Cambrian boundary reminds me a great deal of land animals at another time of mass extinction, the Cretaceous-Tertiary boundary. Beginning near the end of the Triassic Period, dinosaurs dominated terrestrial ecosystems for nearly 150 million years. Mammals shared the landscape for pretty much all of this interval, but remained small and simple. Then, at the Cretaceous-Tertiary boundary, 65 million years ago, a giant meteor slammed into our planet, eliminating the dinosaurs (save for the ecologically distinct birds) but not all mammals. In the ecologically permissive world that followed, mammalian survivors diversified rapidly, giving rise to the

---

[4] The idea of Ediacaran mass extinction didn't originate with me. In 1984, Dolf Seilacher proposed that Ediacaran taxa disappeared many millions of years before the Cambrian began. As John Grotzinger showed, however, this isn't the case. It is the geochemistry of Proterozoic-Cambrian boundary sediments that leads me to think that extinction paved the road to Cambrian diversification.

major groups of mammals that have graced plains and forests ever since. I like the idea that Ediacaran animals were the "dinosaurs" of late Proterozoic ecosystems, simple but ecologically effective organisms that held stem-group bilaterians in check. When Ediacarans vanished, they left a huge opening for bilaterian survivors. Bilaterian animals radiated in this relatively empty world, filling it with the fauna that still dominates the oceans—permissive ecology coming into play once more (figure 12.3).

This argument has been criticized as hard to test. Yes, Ediacaran diversity plummeted at the Proterozoic-Cambrian boundary,[5] but as increased burrowing began to churn sediments in the Cambrian, the preservational window through which we view Ediacaran biology also closed. This isn't quite true, but it is true enough to limit our confidence in interpreting the disappearance of Ediacaran fossils from the rock record. (The Burgess Shale and its earlier Cambrian counterparts do afford an opportunity to look for soft-bodied animals. One or two Ediacaran holdovers have been claimed, but by the end of the Early Cambrian, Ediacaran-grade organisms clearly played a minor role at best in marine ecosystems.)

But, there is another record to test. Remember that microbial reefs in Nama and other late Proterozoic successions contain abundant and modestly diverse skeletons of early animals. Working in Oman, where the C-isotopic boundary event can be traced throughout a large and well-explored basin (it hosts a giant oil field!), John Grotzinger found that *Cloudina*, *Namacalathus*, and other skeletal fossils occur abundantly in all reefs up to the horizon of the isotopic excursion. The microbial reefs continue above it, but they don't contain skeletons. They are almost eerily quiet, much like sedimentary beds just above the Permian-Triassic boundary. Thanks to John, then, we have positive evidence for a biological change of guard precisely coincident with environmental insult at the end of the Proterozoic.

As the Cambrian dawned, there may have been a moment—in a world depleted by extinction and before genetics had locked animals into particular patterns of growth and development—when all things

---

[5] One or two Ediacaran taxa unambigously show up in earliest Cambrian sandstones, but this tells us no more about end-Proterozoic extinction than Triassic brachiopods tell about devastation at the end of the Permian.

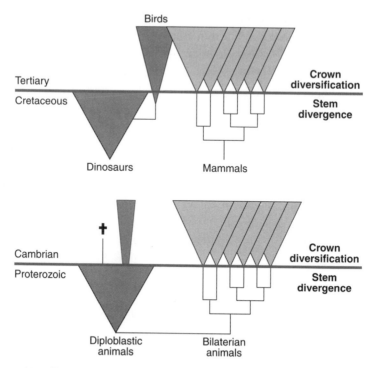

Figure 12.3. Diagram illustrating the suggested analogy between evolution on land at the Cretaceous-Tertiary boundary and evolution in the oceans at the Proterozoic-Cambrian boundary. Diploblastic animals include sponges, cnidarians, and probably, most Ediacaran organisms. (Reprinted with permission from A. H. Knoll and S. Carroll, 1999. Early animal evolution: emerging views from comparative biology and geology. *Science* 284: 2129–2137. Copyright 1999 American Association for the Advancement of Science)

were possible in animal evolution. Any such moment, however, must have been brief. Guided by developmental genetics, expanding animal populations began to accumulate the biological features we associate today with arthropods and brachiopods, echinoderms and chordates. And with emerging body plans came differing functional possibilities that partitioned the metazoan world and shaped the connections among species. Algae diversified, as well, in a Cambrian Explosion that cut across kingdoms.

Physical events may thus have provided the opportunity for Cambrian diversification. But the evolutionary paths actually traveled by Cambrian animals reflect the interplay between development and ecology. Predators and prey locked into an evolutionary arms race, while grazers and algae began to shape the limits of each other's existence. More than ever before, biological interactions and not just the physical environment determined the shape of life. And as the world filled ecologically, evolutionary opportunities for further new body plans dwindled. In the seas, the hand that animal evolution would play for the next 500 million years had been dealt.

Beneath this new ecological edifice, of course, Earth's age-old ecological circuitry continued unchanged. As they did 3 billion years earlier, bacteria continued to cycle biologically important elements through ecosystems, sustaining the biosphere that made animal life possible.

Truth in advertising forces me to acknowledge that some paleontologists tell the story of animal diversification without invoking any environmental influence. In this telling, animals arose only in latest Proterozoic oceans, quickly taking evolutionary shape as the Ediacaran fauna—which, in the predatory world of the Cambrian, gave way to more sophisticated crown group bilaterians and cnidarians. Viewed this way, genetics and ecology are not just important drivers of the Cambrian evolution; they are the only drivers.

The idea that late Proterozoic evolution owes nothing to environmental change has some articulate adherents, but I believe that it flies in the face of everything geology has taught us about both the late Precambrian Earth and more recent evolution. If there is one lesson that paleontology offers to evolutionary biology, other than the documentation of biological history itself, it is that life's opportunities and catastrophes are tied to Earth's environmental history. We can only understand macroevolution—the comings and goings of species and higher taxa through time—if we link the microevolutionary processes studied by geneticists with Earth's dynamic environmental history. The great physical events that framed early animal evolution—global glaciation, the rise of oxygen-filled oceans, and extraordinary perturbations of the carbon cycle—are among our planet's most profound environmental events. We ignore them at our peril.

And here, with the seeds of modern biology planted, my narrative ends. The Cambrian Explosion—both the culmination of life's long Precambrian history and a radical departure from it—emerges as the product of unique interactions between biological and physical processes. The developmental tool kit built by the duplication, mutation, and rearrangement of genes was necessary for animal diversification, but it likely came together at least 600 million years ago and, by itself, could not complete the biological revolution. The new biology required permissive ecology so that unusual genetic variants could survive, and surviving variants to provide the raw material for morphological innovation. In the great physical upheavals that ended the Proterozoic Eon, genetic possibility and environmental opportunity together spawned new and diverse ecosystems in the world's oceans.

More broadly, Earth's long Precambrian history provides illuminating perspective on the great idea of twenty-first-century Earth science—that biology is inexorably linked with tectonics and climate, atmosphere and oceans in a complex and interactive Earth surface system. The pageant of Cambrian evolution simply provides one last, and dramatic, confirmation that life did not evolve on a passive planetary platform. Rather life and environment evolved together, each influencing the other in building the biosphere we inhabit today.

The last word on this history will not be written for many years. At present, we don't really know what processes drove the stepwise growth of oxygen in the atmosphere and oceans. Neither do we understand how and why climates shifted so violently at the beginning and again toward the end of the Proterozoic Eon. And we can only conjecture how these and other events shaped (and were shaped by) biology. We have some leads, some good ideas, and much more data than we had a decade ago—but no resolution. The absence of a definitive punch line may disappoint some readers, but as a paleontologist, it is why I get up in the morning. For scientists, unanswered questions are like Everests unclimbed, an irresistible lure for restless minds.

13 | Paleontology ad Astra

Debate about a small meteorite from Mars has catalyzed scientific interest in one of humanity's oldest questions: are we alone in the universe? What do we know at present; how, in the dawning age of astrobiological exploration, can we learn more; and are there effective limits to what we can learn?

Life arose as a product of physical processes in play on our planet's youthful surface, and it has been sustained for nearly 4 billion years by tectonic, oceanographic, and atmospheric processes that modulate climate and cycle biologically important materials. Perhaps most important, as it has increased and diversified through time, biology has come to provide a suite of planetary processes that is important in its own right. Accepting this planetary view of life on Earth, the next thought is large, if obvious. There are probably a lot of planets out there.

Are we alone in the otherwise sterile vastness of space? Or, might our particular corner of the cosmos be representative of the universe as a whole? To pose such questions is to be human; our grandparents asked them, and their grandparents before them. To find answers, however, may be the special privilege of our own generation.

Of course, the meaning of "we" is critical in this context. If by "we" we mean life, then "we" might well be distributed abundantly across the universe, mostly as simple bacteria-like microorganisms. If, on the other hand, "we" is more narrowly self-referential (life-forms capable of introspection and technology), then "we" might be rare, or even unique. Despite a raft of speculation, we really don't know how to calculate the odds of extraterrestrial life—there is only one data point, and we are it. But speculation is beginning to yield to exploration. I fully expect that one morning in the future, as I sip my coffee in preparation for another day of grateful retirement, I will open the newspaper to find the screaming headline: LIFE FOUND IN SPACE. Again.

On August 7, 1996, I woke up in a hotel room in Beijing, where I was taking part in the International Geological Congress. Still groggy after a fitful night's sleep, I shuffled over to the television in hopes of catching some news before plunging into the day's meetings. As the picture flickered and then flashed into focus, a CNN correspondent, reporting from Baghdad, told the camera (as best I can remember), "I thought I had a big story, but now that they've found life on Mars, my report doesn't seem so important any more." I woke up fast.

By the time I reached the breakfast room, opinions about the announcement were flying thick, fast, and mostly negative: "The fossils are too small." "They got the history of the meteorite all wrong." No one knew any details beyond the few gleaned from television, but everyone had an opinion. Of course, unknown to us, our breakfast in Beijing was duplicating in microcosm an intellectual firestorm that had begun to sweep across the scientific world.

The match that ignited this conflagration was struck by David McKay, a courtly geologist at NASA's Johnson Space Center in Houston, Texas. Along with colleagues at JSC and Stanford University, McKay had reported the discovery of biological fingerprints in a small piece of rock called ALH-84001, a grapefruit-size fragment of Mars blasted into space by bolide impact and delivered to Earth as an unusual and distinctive meteorite (figure 13.1). As announced triumphantly to a credulous press corps, the answer to one of humanity's oldest questions was not only at hand but had, in fact, been sitting quietly in antarctic ice for thousands of years. At least that was NASA's version of the story.

The debate about ALH-84001 has been recounted in countless newspaper stories and a dozen or more books. It has been portrayed both as a courageous leap into the unknown and as a cautionary tale of judgment clouded by desire. The former is certainly true, the latter more arguable. But if David McKay's team *was* misled by an exciting hypothesis, their story resonates because of its universality, not its uniqueness. What scientist has not marched purposefully down a blind alley, buoyed by a great idea that just might be true? Our wings usually get

Figure 13.1. ALH-84001, the grapefruit-size meteorite from Mars that touched off debate about Martian biology. (Photo courtesy of NASA/JPL/Caltech)

clipped in arcane journals and scholarly symposia. Because of its extraordinary public interest, however, the Mars meteorite paper was debated, sometimes acrimoniously, in *Newsweek* and the *New York Times*.

My own early response to the McKay report was neither dramatic nor incisive—I really didn't know what to make of it and thanked the scheduling gods that my trip to China spared me from the frenzy of sound bites unleashed that August. In fact, the opening-night reviews and subsequent pop psychology contain little of lasting value. More impressive are the painstaking analyses that continue to this day, carefully applying color to a decidedly gray area of science. Answers remain elusive, but we now understand that on other planets the familiar questions of life detection must be asked in new ways. As our first, halting exercise in Mars paleontology, ALH-84001 has furnished a planetary extension of the Precambrian research I hold dear. As midwife to the dawning age of astrobiology, it has given us much more.

Before touring the front in the battle over ALH-84001, we should reflect at least briefly on issues that have *not* been contested. First of all, no one argues that ALH-84001 is misidentified as Martian. That, in and of itself, is remarkable. How do we establish the planetary pedigree of a small bit of rock collected from an ice field? Like other meteorites, ALH-84001 has a surface rind of fused minerals that records its heated passage through the atmosphere. In the 1970s, two graduate students named Hap McSween and Ed Stolper proposed a Martian origin for some unusual meteorites called shergottites (after the village of Sherghati, India, where the first known example fell in 1865). To put it politely, their scientific elders were skeptical. More than two decades later, McSween and Stolper are both distinguished scientists, and at least eighteen meteorites are accepted as immigrants from Mars. The most convincing evidence comes from studies of glass in several of these rocks, formed by the impact that blew them into space. The glass contains small inclusions of gas—samples of the atmosphere on the meteorites' parent body. The gas mixture does not resemble terrestrial air, either modern or ancient, but it closely matches the Martian atmosphere as measured by Viking spacecraft. ALH-84001 itself contains no glass that might have captured an atmospheric sample, but is interpreted as Martian on the basis of oxygen locked into its constituent minerals. This oxygen differs in isotopic composition from terrestrial rocks but fits the pattern of other Mars meteorites.

McKay's hypothesis requires that we accept, in principle, two additional suppositions before we pick apart specific points of evidence. First, we must accept that ancient Martians lived in tiny cracks formed within the red planet's crust. Second, we must be willing to believe that these tiny aliens left an interpretable biological record that has survived intact for nearly 4 billion years. This, of course, is where terrestrial experience begins to frame the debate. We find these claims unremarkable because both are met on Earth. Bacteria thrive today thousands of feet beneath the Earth's surface, sustained by chemoautotrophic metabolism in networks of fractures flushed by groundwater. And, as previous chapters make clear, terrestrial rocks preserve an unambiguous signature of early life—admittedly smudged in our oldest examples, but by metamorphism, not age per se. If life ever arose on Mars, it should have left its mark in ancient Martian sediments.

ALH-84001 consists mostly of volcanic rock formed soon after Mars accreted some 4.5 billion years ago. Around 3.9 billion years ago—about the time that life first gained a foothold on Earth—tiny deposits of carbonate minerals formed within cracks that had developed in the rock.

Opinion remains divided on whether the carbonates formed at high temperatures associated with meteorite impact or precipitated from cooler groundwaters that percolated through the cracks in calmer times. Regardless of their origin, the minerals were subsequently modified by transiently high temperatures and pressure imparted by meteorites that continued to pummel the Martian surface. Much later—a mere 16 million years ago—ALH-84001 was ejected into space by one more impact. Captured by Earth's gravitational field, it came to rest 13,000 years ago in the Allan Hills (hence the moniker ALH) of Antarctica.

David McKay's team gleaned four lines of evidence from this meteorite, which collectively convinced them that Mars once supported microbial ecosystems. The carbonate minerals in ALH-84001 (1) resemble terrestrial deposits formed where bacteria are active, (2) contain distinctive grains of the iron oxide mineral magnetite that compare closely to magnetite crystals formed inside bacterial cells, (3) preserve complex organic molecules thought to be derived from biomolecules, and (4) harbor tiny round and rodlike structures interpreted as microfossils.

By themselves, the carbonate minerals tell us only that the cracks in the Allan Hills meteorite once served as a conduit for fluids charged with carbonate and other ions. Like the seafloor precipitates in Archean limestones, they provide unreliable guides to biological processes. The minerals may, however, furnish clues to physical conditions at the time they formed, and our terrestrial experience tells us that environmental setting is key to any paleobiological interpretation. Temperature is of particular interest—life as we know it can persist only where water remains in its liquid form, a few degrees below 0°C in salty, ice-covered ponds of Antarctica to about 113°C in hydrothermal rifts beneath a mile or more of ocean. The isotopic composition of oxygen in the carbonate crystals permits us to estimate temperature, but only if we know the conditions and processes in play when the minerals formed. Geochemists have carefully measured the isotopic composition of ALH-84001, but assumptions about early Mars vary from one computer model to the next, leaving the temperature of carbonate deposition unresolved. If the minerals formed well above the boiling point of water, then biology wasn't present. But if—as many Marsophiles believe—the carbonates formed at cooler temperatures, then organisms could have existed in the cracks of ALH-84001. In neither case, however, do the carbonate crystals *require* biology for their formation.

The organic molecules detected by McKay's team are just as ambiguous. Called polycyclic aromatic hydrocarbons (PAHs, for short), these molecules are well known on Earth. They have been identified in coal and oil, where they were generated by the geologic alteration of originally biological compounds. PAHs also form in industrial furnaces and automobile engines, where organic molecules in coal and gasoline crack and recombine at high temperature. As a result, these distinctive compounds can be found almost ubiquitously in small concentrations. Indeed, PAHs are widespread in the universe, not just on Earth. They have been identified in carbon-bearing meteorites formed by physical processes in the outer solar system, and occur, as well, in interplanetary dust clouds. Like carbonate minerals, therefore, PAHs are not necessarily signs of life.

Simon Clemett and his colleagues at Stanford University have convincingly demonstrated that the minute quantities of PAHs in ALH-84001 are indigenous to the meteorite and not terrestrial contaminants. These scientists do not claim that the PAHs formed from biological precursors—carbonaceous meteorites could easily have introduced the molecules to Mars. Nonetheless, by demonstrating that 4-billion-year-old organic molecules survive intact on our planetary neighbor, Clemett and colleagues have provided information of immense importance to future Mars exploration: if life did arise on Mars, it may have left a molecular calling card in rocks or sediments beneath the Martian surface. We know of no comparably old organic compounds in terrestrial rocks.

What about the tiny structures interpreted as microfossils? The objects in McKay's images (figure 13.2) look like bacterial cells, but this tells us only that they are small and simple. In fact, they are *very* small, less than 100 nanometers long and as little as 20–30 nanometers wide.[1] All bacteria are minuscule, but terrestrial bacteria, compared to the Martian microstructures, are as elephants to mice. At the molecular concentrations found in the common intestinal bacterium *E. coli*, the volume of the structures in ALH-84001 is too small to contain more than a handful of molecules. Does this preclude a biological origin? Not at all.

---

[1] Recall that a nanometer equals one-millionth of a millimeter, or one-thousandth of a micron. The head of a pin is about 1.5 million nanometers long.

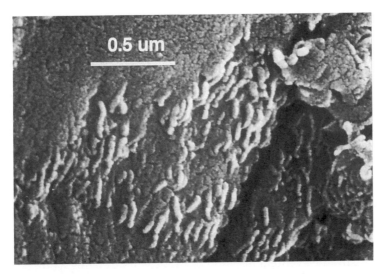

Figure 13.2. Tiny structures interpreted by David McKay and colleagues as nanofossils in ALH-84001. The longest structures are only about 100 nanometers long, not much bigger than the ribosomes found in living cells. (Photo courtesy of NASA/JPL/Caltech)

In the fall of 1998, the National Research Council convened a workshop on the size limits of very small organisms, stimulated in large part by the Mars meteorite debate. Approaching the question from several different starting points, a panel of distinguished cell biologists concluded that it is hard to fit the biochemistry of free-living modern cells into a package smaller than a sphere 200–300 nanometers (0.2 to 0.3 microns) in diameter. In striking concord, a second panel of microbial ecologists reported that in nature free-living cells match but rarely fall below this predicted minimum size. (Viruses are smaller, but they depend completely on the biochemical machinery of their hosts.)

This result has been misinterpreted on both sides of the Mars meteorite debate. It does not indicate that cells smaller than 200–300 nanometers are impossible, only that free-living nanobacteria, if they exist, must have a cell biology of unfamiliar simplicity. Indeed, in the NRC workshop's third panel discussion, molecular biologist Steve Benner proposed that a primitive organism might well fit into a 50-nanometer sphere if it had proteins *or* nucleic acids, but not both (obviating the need for relatively bulky ribosomes to link the two—think RNA world, as in chapter 5). Improbably tiny bacteria have, in fact, been reported from a number of terrestrial environments, including the human bloodstream. These reports have been grasped like life buoys by proponents of Martian (and terrestrial) nanofossils, but have been greeted far more cautiously by most professional microbiologists. Conventional microbiologists (viewed by nanobacteria enthusiasts as a reactionary College of Cardinals to their own Galileo) want evidence that such reported objects are complete and free-living cells.[2] Understandably, they also want plausible molecular explanations for small size. In time, one or more claims of terrestrial nanobacteria may be substantiated, challenging us to rethink our assumptions about contemporary cell biology. But even if this happens, it will not end debate about the Mars microstructures—the observation that cells can be very small does not mean that all very small objects are cells.

Equally, the eventual rejection of nanobacterial claims won't resolve the issue in the other direction. Perhaps the microstructures in ALH-84001 preserve life in its earliest stages—before the evolution of molecular complexity and, thus, unlike modern life on Earth. Or, they might represent conventional cells that shrank during postmortem decay. Or parts of organisms. Size, alone, is inconclusive.

This brings us to the crux of astropaleontological interpretation. We can accept the morphological or chemical patterns in rocks as biological only if they make sense in terms of known biological processes and are unlikely to be made by purely physical mechanisms. That's the rule on Earth—dinosaurs fulfill both criteria, the filamentous microstructures in Warrawoona chert do not—and it is the rule elsewhere in the solar system.

The second criterion is particularly important in planetary exploration. Because we have no assurance that terrestrial organisms exhaust the possibilities of life, we need to think hard about how we might recognize traces of an unfamiliar biology. Extraterrestrial structures or molecules can be accepted as presumptive evidence of biology *only* if we can eliminate the alternative hypothesis that they formed by physical processes. Biology might vary from planet to planet, but physics and

---

[2] The requirement that cells be "free-living" eliminates symbionts or parasites from consideration. Organisms that obtain much of what they need from other cells can jettison many of their genes. I wouldn't be surprised to learn of parasites smaller than known free-living cells.

chemistry should be the same, providing a consistent yardstick for biological assessment.

We don't know the limits of nonbiological pattern formation, and we need to learn them before intelligently chosen samples are returned from Mars. What we do know, however, urges caution. With puckish delight, Spanish geochemist Juan Garcia-Ruiz has concocted chemical mixtures that spontaneously give rise to minute spheres, filaments, and corkscrews of deceptively biological appearance. Garcia-Ruiz's creations do not (yet!) include organic structures comparable to the populations of stalked cells, spiny cysts, multicellular trichomes, and mat-forming colonies described in previous chapters, and for that I am grateful. But neither have such features been identified in ALH-84001. The problem with the Mars microstructures is not that they are too small to be biological, but that they are too simple to exclude abiological interpretations. Indeed, many observers accept a mineralogical origin for the structures. Therefore, by the purposely high standard proposed here, the evocative images published by David McKay do not qualify as evidence of biology on Mars.

As a postscript, Andrew Steele of the Carnegie Geophysical Laboratory *has* found undeniable microorganisms in ALH-84001, but they are terrestrial bacteria. During the long interval when ALH-84001 sat in Antarctica, this dark rock absorbed sunlight to form a small oasis of warmth in the polar desert. Bacteria took refuge in cracks once washed by Martian water—undeniable testimony to the ubiquity and hardiness of life on Earth, but yet another roadblock in the search for *Martian* biosignals.

Of the four lines of evidence originally proposed by the McKay team, only one has survived into the new millennium, the unusual magnetite crystals in ALH-84001's carbonate globules. The magnetic properties of magnetite are well known, but its biological associations are not. In fact, magnetite crystals can provide evidence of biology *because* of their magnetism—some bacteria sense direction using a chain of elongated magnetite crystals synthesized within their cells. The tiny grains have a crystal form and chemical purity not found in magnetites from igneous or metamorphic rocks. After cells die, bacterial magnetite can be deposited in sediments, and magnetofossils have been found in rocks as old as 2.0 billion years. On Earth, then, if you find a really pure magnetite in nature, it's a sign of life.

Magnetite grains in ALH-84001 come in various shapes and sizes. Some have structural defects in their crystals that preclude a biological origin. A subset, however, display mineralogical features associated on Earth with biology. Joe Kirschvink, introduced earlier as father of the Snowball Earth hypothesis, believes that similar crystals require similar explanations. To Joe and the McKay team, the magnetite grains in ALH-84001 provide a smoking gun for Martian biology. They argue that the industrial importance (think magnetic tapes) of tiny, chemically pure magnetite crystals is such that if physical synthesis were possible, some enterprising chemist would already have discovered how to do it. Possibly so, but this argument can be stood on its head. Perhaps the person who penetrates the secrets of Martian magnetite will land a patent of immense commercial value.

And so it may. Earlier, I noted that while the carbonates in ALH-84001 may have formed at mild temperatures, they subsequently experienced transiently high temperature and pressure when bombarded by meteorites. With this in mind, a team of mineralogists from the Johnson Space Center and nearby engineering firms devised a most revealing laboratory experiment. Mixing sodium bicarbonate and iron, calcium, and magnesium salts in CO_2-charged water (approximating the chemical composition of the precipitated globules in ALH-84001), the team induced mineral precipitation at relatively low (150°C) temperatures. Then they subjected the deposited minerals to a simulated meteorite impact, sharply increasing the pressure and temperature (to 470°C) of the reaction vessel for a short time. When the mixture cooled, the team found chemically pure, defect-free magnetite crystals much like those in ALH-84001. Moreover, these newly synthesized magnetites display the distinctive crystal form associated on Earth with biology.

The great twentieth-century philosopher Karl Popper famously maintained that scientific hypotheses can never be proved, only disproved—one thousand black swans can't prove the hypothesis that all swans are black, but a single white swan can show it to be wrong. This seems straightforward enough, but as the *New Yorker*'s Adam Gopnik has observed, real scientific debates are seldom so tidy. Confronted by a white swan, advocates of the black swan hypothesis are likely to question the evidence—"You call that a swan?" in Gopnik's droll telling. So it goes with Mars magnetite. Those who see biology in Martian magnetite vigorously dispute the claim that the loyal opposition has produced anything like the crystals found in ALH-84001.

Regardless of how this disagreement turns out, one more line of evidence challenges the notion that life played a hand in forming Martian magnetite crystals. Based on careful electron microscopic and X-ray microstructural studies, British scientists David Barber and Edward Scott have confirmed earlier observations that the crystal structures and orientations of magnetite grains in the Allan Hills meteorite reflect the crystallographic properties of the carbonates that surround them— something we might predict if the magnetites formed during meteor impact, but unlikely if biology grew the crystals.

Debate about martian magnetite continues, but support for a biological origin is waning, and with it, the last hope that questions of extraterrestrial life might be solved easily. Thinking back to the Proterozoic rocks of Spitsbergen, Gunflint, and the Great Wall, we have to recognize that *none* of the compelling biosignatures found in those rocks have been identified in ALH-84001. People argue about the crystal structure of Martian magnetite because there are no unambiguous microfossils, no steranes, and no microbial mat structures in ALH-84001 or any other meteorite from Mars.

Where does all this leave us? The cynic might answer that where fossils are absent, it's hard to detect ancient life. A more forward-looking conclusion is that if we want to know whether Mars has ever been a biological planet, we'll have to go there. And before we go, we've got some homework to do.

If ALH-84001 nurtured the discipline of astrobiology in its infancy, what will become of this field in maturity? One dividend of astrobiological thinking, already apparent, is the growth of a planetary perspective on terrestrial biology. Microbial ecologists, biogeochemists, paleontologists—all of us now routinely consider the ramifications of new discoveries for our understanding of the Earth system as a whole. But an astrobiology that truly meets the challenge of its name cannot remain Earthbound. It requires exploration, and the entire universe beckons.

Whatever one might conclude from the meteorite debate, Mars

remains our prime target for astrobiological investigation, and not just because of its proximity. The present-day Martian landscape is bleak beyond imagination, hardly a place to look for life. But channeled terrains and other surface features preserved over 4 billion years tell us that early in its history, Mars was much more like Earth (figure 13.3). Both planets had relatively thick atmospheres, active volcanism, and, at least intermittently, liquid water. On Earth, these conditions incubated life, and it is reasonable to hypothesize that they could also have done so on Mars.

One school of thought, championed with particular eloquence by astrophysicist Paul Davies, holds that Mars actually beat Earth to the mark. Davies believes that life may have begun on Mars and colonized the Earth by traveling in meteoritic "spacecraft." ALH-84001 and its kin show us that the required trade route exists, and it is indeed possible that simple organisms in meteorite interiors could have survived launch, prolonged exposure to radiation in space, and arrival on Earth. The biggest hurdle might well have been ecological—what is the likelihood that a meteorite would land on a substrate its microscopic occupants could metabolize? The "we are all Martians" hypothesis is imaginative, as well as a bit subversive. It subtly undermines the notion that life might arise wherever Earth-like planets exist, because if it is correct, life didn't originate on Earth. Before worrying about this overmuch, however, we should probably learn whether Mars ever supported biology.

Bruce Jakosky, a planetary scientist at the University of Colorado, and I once spent an idle afternoon concocting a title for a brief essay about Mars astrobiology. After (too?) much discussion, we arrived at "The Search for Life on Mars: Fish in a Barrel, Needle in a Haystack, or Pig in a Poke?" We chuckled over our cleverness for weeks afterward, but, perhaps wisely, never published the essay. After that title, there wasn't much else to say.

We can readily eliminate one possibility raised in that tongue-in-cheek title—looking for life on Mars won't be like shooting fish in a barrel. By the most optimistic estimates, we will set virtual foot on Mars a half dozen times in the next twenty years. Given NASA's current ban on launching nuclear power sources, landing missions will be relatively short and the exploration radii consequently small. (On an early field test of the 2003 Mars rover conducted in a forlorn valley not far from Las

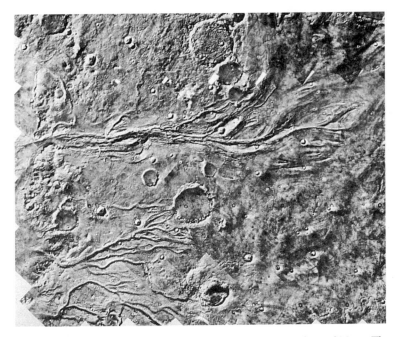

Figure 13.3. Channel network carved by water into the surface of Mars. These channels formed early in the red planet's history, principally by groundwater sapping and catastrophic melting. How much liquid water existed at any one time—whether early Mars had persistent oceans or long-lived rivers and lakes—remains a subject of debate. (Viking Image courtesy of NASA/JPL/Caltech)

Vegas, I wandered over to a nearby outcrop, looked it up and down, chipped off a bit of rock with my hammer, and examined it briefly by hand lens. Satisfied with my observations, I discarded the rock chip and walked back to the group. "That took three minutes," noted Cornell's Steve Squyres, principal investigator for the project. "On Mars, the rover will need a full day to do the same thing.")

Geologists interested in Mars's planetary history favor landing sites that resemble the Grand Canyon; engineers charged with landing safely prefer Kansas. Given that we can explore only a small fraction of the Martian landscape by lander, there is a premium on learning all we can by orbital observation in order to choose sites that optimize scientific opportunity and technical feasibility. Our experience in Precambrian paleontology tells us where to look and how—promising targets include precipitated spring deposits and fine mudstones formed beneath ancient water bodies. But, as noted above, terrestrial experience may not prepare us for what we find—our biological and paleobiological knowledge of terrestrial life furnishes blinders as well as a guide. Opportunities for false positives and false negatives abound, and it will be a challenge to get it right in the few precious opportunities we are granted.

We have no guarantee that life ever existed on Mars, and the aggressively oxidizing nature of its present surface will undoubtedly have erased some evidence that may once have lain within the grasp of rovers. Even if we succeed in coring sedimentary rocks to obtain unaltered samples of ancient shales, astropaleontology may prove frustrating. On Earth, the fingerprints of biology are everywhere because of photosynthesis, which enabled life to expand across our planet's surface. In the absence of photosynthetic organisms, traces of Martian life (if it existed) might be common only in the immediate vicinity of hydrothermal springs.

There is, of course, another possibility, and that is to search for living and not fossil organisms. For years a small band of optimists has reasoned that chemosynthetic microorganisms might persist today in hydrothermal oases far below the Martian surface. It's a long shot, to be sure, but in a world without data, all things are possible and most are worth investigating. It may be possible to detect subsurface water by microwave imaging from orbiting satellites, but at present deep drilling is well beyond our capabilities. For this reason, astrobiologists were energized by the recent announcement that liquid water has been present on the Martian surface within the past few million years and conceivably could be there today. The evidence, recognized by planetary scientists Michael Malin and Kenneth Edgett in images from the *Mars Global Surveyor*, consists of erosional gullies that originate in crater and valley walls and cut into the aprons of talus at their bases (figure 13.4). The gullied talus slopes contain few craters, a sign of relative youth among Martian landforms. Some gullies cut across dunes, but no dunes obscure gullies, reinforcing the view that these gullies are no more than a few million years old. On Earth, water sculpts similar features, and so Malin and Edgett favor groundwater seepage and surface runoff as an explanation for the Martian gullies. Quite a few planetary scientists have welcomed this interpretation, although Michael Carr, the doyen of Martian water, cautions that other explanations may be possible. As Carr reminds us, the gullies occur at mid- to high latitudes where the

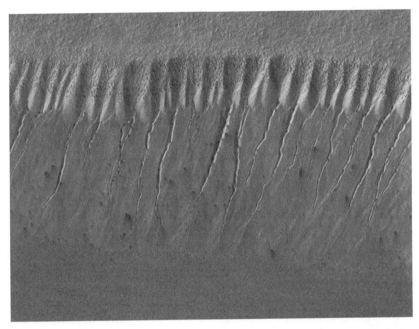

Figure 13.4. Gullies cut in relatively recent times into talus slopes beneath the walls of Martian craters. Such features suggest that liquid water can still form at the Martian surface—but how continuously it forms and how long it lasts remain uncertain. (Mars Global Surveyor Image courtesy of NASA/JPL/Malin Space Systems)

Martian surface is so cold that the presence of liquid water is hard to understand. One possibility is that Mars's axis of rotation swings widely through time. This means that features currently in a deep freeze might thaw every few million years. That idea is intriguing but not necessarily cheering to astrobiologists—environments where liquid water is present for 1 million years and then absent for the next don't easily sustain biology.

The newly discovered Martian gullies present an exciting research challenge that promises to tell us much about Mars as a planet, regardless of what it reveals about habitability. Landing on them, however, won't be easy—in this instance the "Grand Canyon problem" looms large.

My thoughts on Mars astrobiology emphasize problems over opportunities—a needle-in-a-haystack view, if you will. Am I too skeptical? Maybe, but naive enthusiasm has pervaded so much Mars rhetoric that a mild corrective seems in order. We must not assume that the astrobiological exploration of Mars will be easy or that unambiguous answers will soon be in hand. The search for life on Mars will be difficult and may turn up nothing, but as part of a balanced effort to understand our planetary neighbor, it is worth doing and worth doing right. Bear in mind that a negative conclusion to our search will be just as important as a positive answer. If a thick atmosphere, volcanism, and liquid water do *not* provide a foolproof recipe for life, we'll have to rethink our models of biogenesis—while feeling a bit lonelier than before.

The new age of Mars exploration is just beginning, and it is likely to continue well beyond my lifetime. No knows what we will find. Perhaps, then, we should leave the final word to a poet, at least for now. In his great novel *The Glass Bead Game*, Hermann Hesse wrote two beautiful lines that provide a fitting credo for Mars astrobiology:

> Nothing is harder, yet nothing is more necessary, than to speak of certain things whose existence is neither probable nor demonstrable. The very fact that serious and conscientious men treat them as existing things brings them a step closer to existence and to the possibility of being born.

Mars may be a particularly attractive target for astrobiological exploration, but it is hardly the only one. Within our solar system, there is Europa, a Jovian moon whose icy surface may cover an ocean of water, and Titan, a moon of Saturn with an atmosphere of methane and hydrocarbon smog. Farther afield, the targets are potentially endless, but as we leave our solar system the rules of the game change. Some day, our descendants may find a loophole in the laws of physics, but until they do, astrobiological exploration of the greater universe will be done remotely.

One of the most exciting recent developments in astronomy has been the discovery of planets in orbit around nearby stars. So far, we've observed only giant planets (about the size of Saturn, or larger) in tight orbits around their stars, but that isn't necessarily because such planets dominate the universe. At present, these are the only planets we can detect. By the end of the next decade, however, an ambitious project called the Terrestrial Planet Finder may enable us to detect Earthlike planets in nearby solar systems. Not only that, spectroscopic images may reveal the compositions of their atmospheres.

In the early 1970s, James Lovelock, father of the Gaia hypothesis, proposed that planetary atmospheres provide sensitive indicators of biological activity. To appreciate why, we need only look at the Earth. In addition to nitrogen gas and carbon dioxide, our atmosphere contains water vapor, oxygen, and a small but measurable amount of methane. It isn't easy to see how oxygen and methane could coexist in an atmosphere governed by equilibrium chemical processes, but when cyanobacteria (or their chloroplast descendants) and methanogens are afoot, this mixture can be sustained indefinitely. (Of course, by this methodology, we couldn't have pegged the Earth as biological for the first half of its history.)

We will not soon learn whether the inhabitants of extrasolar planets synthesize DNA or proteins, whether they are unicellular or include many-celled forms, or whether they live on land as well as in water. But on a distant planet where organisms are abundant and include the right metabolisms, we might recognize life by its environmental impact.

At distances greater than a few light years, however, even planet finders will fail us—planets in more distant galaxies are simply too dim and too far away to be detected in this way. For most of the vastness of the universe, then, we can search for life in only one way—by listening for technologically gifted beings who can answer our signals or initiate dialogue. On Earth, such capabilities have become available only within the past century.

In their book *Rare Earth*, Peter Ward and Donald Brownlee argue that intelligent life may be exceedingly uncommon in the universe, citing the myriad of astronomical and tectonic circumstances that contributed to the evolution of neurological complexity on Earth. Hap McSween earlier presented similar arguments in his fine volume *Fanfare for Earth*. Sometimes lampooned as "Goldilocks" hypotheses because they require everything to be "just right" for the evolution of intelligence, these arguments assume that because the conditions that facilitated our own evolution are particular, they must be rare. But we have no way of knowing whether this is true. We can agree that not all solar systems contain planets that are Earthlike in every way, but what if 10 percent do, or 1 percent, or even one in a million. Given the dimensions of the universe, this would provide untold millions of potential incubators for intelligent life. Put another way, planets with intelligent life could be proportionally rare but absolutely abundant. We really don't know how to assess the odds. And there is another kicker: we have no idea whether the route by which we achieved intelligence is the only one available. I doubt that it is, but I can't come up with a convincing way to couch my prejudices in theory. The issue must be solved empirically.

It is ironic that the type of life most likely to be rare is the only one we can search for through most of the universe. It is also unfortunate, because the vastness of space makes it difficult and perhaps impossible to address one of astrobiology's most fundamental questions: which attributes of terrestrial biology are general features of life and which are local consequences of our own peculiar history? If we don't find life on Mars, and if we fail to detect organisms in Europan ice or oceans, we may never learn the answer—unless some alien with a crystal set tells us.

Epilogue

> The past is autobiographical fiction pretending to be
> a parliamentary report.
> —Julian Barnes
> *Flaubert's Parrot*

"How do we seize the past?" muses the narrator in *Flaubert's Parrot*. "We read, we learn, we ask, we remember, we are humble; and then a casual detail shifts everything." And so it does. In our efforts to decipher Earth's early biological history, minute details have repeatedly sparked revelation: here an exquisite cyanobacterium in billion-year-old chert, there a drop in the iron retained by ancient soils, and somewhere else a change in the fabric of carbonate minerals precipitated on the ocean floor—the world in a grain of sand if ever there was one. Individual details may seem trifling, but collectively they reveal the epic drama that carries us from Darwin's warm little pond to his dilemma of Cambrian diversity, now seen in new light. From there, it is but a short hop to fish, to ungainly amphibians waddling across a Paleozoic swamp, to tiny mammals dodging the footfalls of dinosaurs, and to a species that can not only reconstruct its evolutionary past but contemplate similar histories on other worlds.

The early evolution of life, then, is part of *our* story. We are the product of a planetary history more than 4 billion years long, the latest installment of a book whose final chapters have yet to be written. As Hap McSween wrote in *Fanfare for Earth*, "we are stardust" is not just Woodstock bravado; it is literal truth. The carbon in my body was forged in the crucible of an early star, dispersed into space by a supernova, gathered along with dust and rock as our planet took shape, and then cycled repeatedly among air, oceans, and organisms, through cyanobacteria and dinosaurs, perhaps even through Darwin, before coming to rest, at least for the moment, in a paleontologist's brain.

But while the story of evolution undoubtedly includes human beings, it is not *about* us. The long history of life helps to explain our presence, but it can be interpreted as a journey toward man only if we strike a particular course through the Tree of Life. Travel another path and life's history is a gripping saga of cyanobacterial survival, a cautionary tale of trilobitic fall, or the inspirational story of yeasts finding sustenance in rotting fruit. Each of the 10 million or so species alive today is *equally* the product of Earth's 4-billion-year evolutionary history—myriad forms separated by evolutionary divergence but united in ecological codependence. Whatever the merits of viewing Earth as *our* world, we could not persist without the bacteria and algae, plants and animals. We are evolutionary latecomers, among the latest threads in an ecological tapestry woven since our planet was young.

It is, in fact, ecology that confers special status on humans, not evolution. Unlike the millions of species that preceded us, humans don't simply adapt to the environments provided by nature. We take our environment with us, finding comfort in a heated Siberian cabin or an air-conditioned condo in Houston. Armed with technology, our species has spread across the planet, populating it in remarkable numbers. And in the process, we have altered nearly all landscapes, commandeered much of Earth's photosynthetic production, and come to rival bacteria as participants in biogeochemical cycles. What we do with our special status will determine the plot of the next chapters in Earth's history, not just for us but for the biosphere as a whole.

There are, of course, other versions of the tale. And not just the "autobiographical fictions" of other paleontologists who refract the same observations through a different lens of experience. I'm thinking of versions that slip free of the factual moorings that guide and constrain my telling—granting us special status by assertion rather than ecology, while simultaneously rejecting almost everything else argued in this book. How do we think about explanations of Earth and life that dispense with science altogether?

The great creation stories of the Bible, or the Upanishads, or the Aboriginal Dreamtime provided ways of comprehending the universe thousands of years before Copernicus, Newton, Darwin, and Einstein furnished new explanatory language. As eloquent guides to a *moral* universe, they continue to speak across the generations. Indeed, their power derives from their timelessness—words that inspired an iron age shepherd in the Levant can still move a computer analyst in Detroit. Scientific accounts, in contrast, are bounded in time. Today's state of the art was incomprehensible yesterday, and it will be out-of-date tomorrow. That these two ways of comprehending should be confused in either form or purpose strikes me as both absurd and unfortunate.

The modern world provides substantial tests for faith and theology—the Holocaust, crib death, and Alzheimer's disease come readily to mind. In contrast, the reconciliation of traditional truths and science is almost trivially simple, requiring only that God, if present, be great enough to mix immanence into the nascent universe, enabling it to unfold over the eons, obedient to the laws of special relativity, nuclear chemistry, and population genetics. Science's creation story accounts for process and history, ✓ not intent. Accepting its ancient counterparts as parables, then, eliminates conflict. (Saint Augustine said as much in the fourth century.) But we must be clear: there can be no other resolution that involves science.

Creationists commonly target evolutionary biology as science's boogeyman, but the account of early evolution presented in preceding chapters necessitates that the biblical literalist be catholic in his rejection of scientific understanding. He must reject geology because its confluence of pattern and process cannot be accommodated by a biblical timetable. Physics and chemistry must go, too, because they explain the radioactive decay that dates zircons as millions or billions of years old. And astronomy and astrophysics? Don't even think about them. Indeed, the biblical literalist, passing Permian brachiopods, Cambrian trilobites, and 1.7-billion-year-old schist as she hikes down the Grand Canyon, can only conclude that the *appearance* of age and order in stratigraphic successions is an elaborate ruse, part of a great cosmic charade set up to trap the unfaithful. What sort of God would do that? One who can be petty and vengeful, who may love His creation but doesn't trust it. A God, in other words, much like ourselves. In his zeal to know the mind of God, the creationist finds only a mirror.

Of course, scripture insists that God made man in his own image, not the reverse. To a nomad seeking oases in a Mideastern desert or a seamstress laboring in medieval Europe, this may well have been received as a literal commentary on God's visage. Philosophers from Aquinas to

Descartes saw God reflected in the human mind. But the scientific and technological revolution of the twentieth century suggests a more specific, and perhaps more unsettling, reading. To a remarkable degree, we have come to understand the world we live in and, indeed, to dominate it. Through physics and engineering, we can harness the power of atoms for electricity or mass destruction. Medicine makes lame beggars walk. We can fathom the miracle of birth and the mystery of death, and have the power of life and death over species as well as people. Perhaps we were made in God's image after all.

In the end, dialogue between religion and science matters not so much because it holds the prospect of consensus on our past, but because we need to agree about our future. At the dawn of the twenty-first century, we stand at a crossroads in Earth history. The technological intelligence that gained ecological hegemony for humans now threatens the products of a planetary lifetime. As a result, our grandchildren may know the rhinoceros only from pictures, the rain forest from parks, and coral reefs from history books. Even as we search for life on Mars, we risk losing it on Earth.

Thoughts such as these are disheartening. But the future needn't be an evolutionary endgame. There is another possibility. At the intersection of ecological dominance and planetary history lie the makings of an evolutionary ethics. If we can understand the immensity of our evolutionary inheritance, we may be moved to preserve it. If we can acknowledge our unprecedented role as planetary stewards, we may be able to discharge our responsibility with wisdom and with honor. On this issue, at least, faith and science find common ground. I don't know whether God decreed the passenger pigeon, but if He did, it was not for us to exterminate.

Copernicus and Darwin profoundly altered the human sense of self. We do not live at the center of the universe, and we cannot claim the privileges of special creation. In coming decades, planetary exploration may even show that we are not unique or, at the very least, not alone. But whatever astronomy and evolution may take away, ecology restores. On this planet, at this moment in time, human beings reign. Regardless of who or what penned earlier chapters in the history of life, we will write the next one. Through our actions or inaction, we decide the world that our grandchildren and great grandchildren will know. Let us have the grace and humility to choose well.

Further Reading

Prologue

Whitman, W. 1993. When I heard the learn'd astronomer, p. 340 in *Leaves of Grass*. Reprint of the "Deathbed Edition," originally published in 1892. Modern Library, New York.

Chapter 1. In the Beginning?

Key References on Kotuikan Geology and Paleontology

Bowring, S. A., J. P. Grotzinger, C. E. Isachsen, A. H. Knoll, S. M. Pelechaty, and P. Kolosov. 1993. Calibrating rates of Early Cambrian evolution. *Science* 261: 1293–1298.

Kaufman, A. J., A. H. Knoll, M. A. Semikhatov, J. P. Grotzinger, S. B. Jacobsen, and W. Adams. 1996. Integrated chronostratigraphy of Proterozoic-Cambrian boundary beds in the western Anabar region, northern Siberia. *Geological Magazine* 133: 509–533.

Khomentovsky, V. V., and G. A. Karlova. 1993. Biostratigraphy of the Vendian-Cambrian beds and the lower Cambrian boundary in Siberia. *Geological Magazine* 130: 29–45.

Rozanov, A. Yu. 1984. The Precambrian/Cambrian boundary in Siberia. *Episodes* 7: 20–24.

Selected General References

Barnes, J. 1986. *Staring at the Sun*. Jonathan Cape, London. (Source of my opening quotation; reprinted with permission.)

Conway Morris, S. 1998. *The Crucible of Creation: The Burgess Shale and the Rise of Animals*. Oxford University Press, Oxford. (An individualistic but authoritative account of the Cambrian Explosion.)

Darwin, C. 1859. *On the Origin of Species by Means of Natural Selection*. J. Murray, London. (Often reprinted, Darwin's masterpiece is the foundation of modern biology.)

Fortey, R. 1996. *Life: A Natural History of the First Four Billion Years of Life on Earth*. Alfred Knopf, New York. (A good introduction to paleontology and paleontologists, with much to say about early animal evolution.)

Gould, S. J., and N. Eldredge. 1993. Punctuated equilibrium comes of age. *Nature* 366: 223–227. (Commentary on the idea of punctuated equilibrium and the relationship of stratigraphic pattern to evolutionary process.)

Chapter 2. The Tree of Life

Bult, C. L., and 40 others. 1996. Complete genome sequence of the methanogenic archaeon *Methanococcus janaschii*. *Science* 273: 1058–1073. (Among the first microbial genomes to be published, this paper established beyond doubt the distinctive nature of archaeal biology.)

Doolittle, W. F. 1994. Tempo, mode, the progenote, and the universal ancestor. *Proceedings of the National Academy of Sciences, USA* 91: 6721–6728. (A lively review of efforts by molecular biologists to root of the Tree of Life using duplicated genes.)

Doolittle, W. F. 2000. Uprooting the Tree of Life. *Scientific American* 282 (2): 90–95. (A discussion of gene trees, organismsic phylogeny, and the differences between them in evolutionary investigations of microorganisms.)

Fitz-Gibbon, S. T., and C. H. House. 1999. Whole genome-based phylogenetic analysis of free-living microorganisms. *Nucleic Acids Research* 27: 4218–4222. (Presents a phylogeny based on whole-genome analysis that compares closely with trees based on ribosomal RNA sequences.)

Madigan, M. T., J. M. Martinko, and J. Parker. 1999. *Brock Biology of Microorganisms*, eighth edition. Prentice Hall, New York. (A good place to learn about the biology and diversity of Bacteria and Archaea.)

Miller, R. V. 1998. Bacterial gene swapping in nature. *Scientific American* 278 (1): 67–71. (A primer on the lateral transfer of genes by bacteria.)

Nealson, K. H. 1997. Sediment bacteria: Who's there, what are they doing, and what's new? *Annual Review of Earth and Planetary Science* 25: 403–434. (A superb introduction to microbial metabolism and ecology.)

Ochman, H., J. G. Lawrence, and E. A. Grossman. 2000. Lateral gene transfer and the nature of bacterial innovation. *Nature* 405: 299–304. (An up-to-date review of our growing knowledge of the role played by lateral gene transfer in bacterial evolution, more advanced treatment than Miller 1998.)

Pace, N. R. 1997. A molecular view of microbial diversity and the biosphere. *Science* 276: 734–740. (A fine summary of how molecular biology has reshaped our understanding of microbial evolution and ecology.)

Stetter, K. O. 1996. Hyperthermophiles in the history of life. *Ciba Foundation Symposium* 202: 1–18. (Review of archaeal diversity and ecology by one of the pioneers in research on these organisms.)

Woese, C. R. 1987. Bacterial evolution. *Microbiological Reviews* 51: 221–271. (A comprehensive summary of Woese's pioneering views on bacterial phylogeny, based on sequence comparison of genes for small-subunit ribosomal RNA.)

Chapter 3. Life's Signature in Ancient Rocks

Key References to the Precambrian Paleontology of Spitsbergen

Butterfield, N. J., A. H. Knoll, and K. Swett. 1994. Paleobiology of the Neoproterozoic Svanbergfjellet Formation, Spitsbergen. *Fossils and Strata* 34: 1–84.

Harland, W. B. 1997. *The Geology of Svalbard*. Geological Society Memoir 17, 521 pp.

Knoll, A. H., J. M. Hayes, A. J. Kaufman, K. Swett, and I. B. Lambert. 1986. Secular variation in carbon isotopic ratios from upper Proterozoic successions of Svalbard and East Greenland. *Nature* 321: 832–838.

Knoll, A. H., and K. Swett. 1990. Carbonate deposition during the late Precambrian era: An example from Spitsbergen. *American Journal of Science* 290A: 104–131.

Knoll, A. H., K. Swett, and J. Mark. 1991. Paleobiology of a Neoproterozoic tidal flat/lagoonal complex: The Draken Conglomerate Formation, Spitsbergen. *Journal of Paleontology* 65: 531–570.

Selected General References on Precambrian Paleobiology

Des Marais, D. J. 1997. Isotopic evolution of the biogeochemical carbon cycle during the Proterozoic eon. *Organic Geochemistry* 27: 185–193. (A demanding but rewarding essay on the use of isotopic data to reconstruct ancient biogeochemical systems.)

Grotzinger, J. P., and A. H. Knoll. 1999. Precambrian stromatolites: Evolutionary milestones or environmental dipsticks? *Annual Review of Earth and Planetary Sciences* 27: 313–358. (Detailed arguments on how to interpret ancient stromatolites.)

Knoll, A. H. 1996. Archean and Proterozoic paleontology, pp. 51–80 in J. Jansonius and D. C. MacGregor, editors, *Palynology: Principles and Applications*, volume I. American Association of Stratigraphic Palynologists Press, Tulsa. (An up-to-date, well-illustrated review of Precambrian microfossils.)

Knoll, A. H., and D. E. Canfield. 1998. Isotopic inferences on early ecosystems. *The Paleontological Society Papers* 4: 211–243. (A primer on the integration of isotopic, paleontological, and phylogenetic information in Precambrian research.)

Schopf, J. W. 1999. *Cradle of Life*. Princeton University Press, Princeton, N.J. (A first-person account of the early development of Precambrian paleontology.)

Schopf, J. W., and C. Klein, editors. 1992. *The Proterozoic Biosphere: A Multidisciplinary Study*. Cambridge University Press, Cambridge. (A massive and authoritative, if, by now, somewhat dated review covering all aspects of Precambrian paleobiology.)

Summons, R. E., and M. R. Walter. 1990. Molecular fossils and microfossils of prokaryotes and protists from Proterozoic sediments. *American Journal of Science* 290A: 212–244. (An accessible and expert introduction to biomarkers in Precambrian sedimentary rocks.)

Walter, M. R., editor. 1976. *Stromatolites*. Elsevier, Amsterdam. (More than two decades old, but still the Bible of Precambrian stromatolite studies.)

Chapter 4. The Earliest Glimmers of Life

Key References on Warrawoona Geology and Paleobiology

Barley, M. E., and S. E. Loader, editors. 1998. The tectonic and metallogenic evolution of the Pilbara Craton. *Precambrian Research* 88: 1–265. (A compendium of tectonic and geochronological data on the Warrawoona Group and related rocks, including the paper by Nijman and colleagues on barite mounds and other features of Warrawoona geology.)

Brasier, M. D., O. R. Green, A. P. Jephcoat, A. K. Kleppe, M. J. van Kranendonk, J. F. Lindsay, A. Steele, and N. V. Grassineau. 2002. Questioning the evidence for Earth's oldest fossils. *Nature* 416: 76–81.

Buick, R., J.S.R. Dunlop, and D. I. Groves. 1983. Stromatolite recognition in ancient rocks: An appraisal of irregularly laminated structures in an early Archaean chert-barite unit from North Pole, Western Australia. *Alcheringa* 5: 161–181.

Buick, R., J. R. Thornett, N. J. McNaughton, J. B. Smith, M. E. Barley, and M. Savage. 1996. Record of emergent continental crust ~3.5 billion years ago in the Pilbara Craton of Australia. *Nature* 375: 574–577.

Groves, D. I., J.S.R. Dunlop, and R. Buick. 1981. An early habitat of life. *Scientific American* 245 (10): 64–73.

Hofmann, H. J., K. Grey, A. H. Hickman, and R. I. Thorpe. 1999. Origin of 3.45 Ga coniform stromatolites in Warrawoona Group, Western Australia. *Geological Society of America Bulletin* 111: 1256–1262.

Kerr, R. A. 2002. Reversals reveal pitfalls in spotting ancient and E.T. life. *Science* 296: 1384–1385.

Lowe, D. R. 1983. Restricted shallow water sedimentation of early Archaean stromatolitic and evaporitic strata of the Strelley Pool chert, Pilbara Block, Western Australia. *Precambrian Research* 19: 239–248.

Lowe, D. R. 1994. Abiological origin of described stromatolites older than 3.2. Ga. *Geology* 22: 387–390.

Schopf, J. W. 1993. Microfossils of the early Archean Apex Chert: New evidence of the antiquity of life. *Science* 260: 640–646.

Schopf, J. W., A. B. Kudryavtsev, D. G. Agresti, T. Wdowiak, and A. D. Czaja. 2002. Laser-Raman imagery of Earth's earliest fossils. *Nature* 416: 73–76.

Schopf, J. W., and B. Packer. 1987. Early Archean (3.3-billion to 3.5-billion-year-old) microfossils from Warrawoona Group, Australia. *Science* 237: 70–73.

Shen, Y., D. Canfield, and R. Buick. 2001. Isotopic evidence for microbial sulphate reduction in the early Archaean ocean. *Nature* 410: 77–81.

Walter, M. R., R. Buick, and J.S.R. Dunlop. 1980. Stromatolites 3,400–3,500 Myr old from the North Pole area, Western Australia. *Nature* 284: 443–445.

Key References on Barberton Paleobiology

Byerly, G. R., D. R. Low, and M. M. Walsh. 1986. Stromatolites from the 3,300–3,500-Myr Swaziland Supergroup, Barberton Mountain Land, South Africa. *Nature* 319: 489–491.

Knoll, A. H., and E. S. Barghoorn. 1977. Archean microfossils showing cell division from the Swaziland System of South Africa. *Science* 198: 396–398.

Lowe, D. R., and G. R. Byerly, editors. 1999. Geological evolution of the Barberton Greenstone Belt, South Africa. *Geological Society of America Special Paper* 329.

Walsh, M. M. 1992. Microfossils and possible microfossils from the early Archean Onverwacht Group, Barberton Mountain Land, South Africa. *Precambrian Research* 54: 271–293.

Walsh, M. M., and D. R. Lowe. 1999. Modes of accumulation of carbonaceous matter in the early Archean: A petrographic and geochemical study of the carbonaceous cherts of the Swaziland Supergroup. *Geological Society of America Special Paper* 329: 115–132.

Westall, F., M. J. de Wit, J. Dann, S. van der Gaast, C.E.J. de Ronde, and D. Gerneke. 2001. Early Archean fossil bacteria and biofilms in hydrothermally influenced sediments from the Barberton greenstone belt, South Africa. *Precambrian Research* 106: 93–116.

Selected General References on the Early Archean Earth

Bowring, S. A., and T. Housh. 1995. The Earth's early evolution. *Science* 269: 1535–1540. (A summary of how recently acquired geochemical data are changing views about early Earth history.)

Fedo, C. M., and M. J. Whitehouse. 2002. Metasomatic origin of quartz-pyroxene rock, Akilia, Greenland, and implications for Earth's earliest life. *Science* 296: 1448–1452. (Argues that putative biosignatures claimed in Akilia rocks by Mojzsis and colleagues—see below—actually originated by physical processes during metamorphism.)

Kasting, J. F. 1993. Earth's early atmosphere. *Science* 259: 920–926. (A good review that uses geochemical data and atmospheric modeling to draw inferences about Archean air.)

Mojzsis, S. J., G. Arrhenius, K. D. McKeegan, T. M. Harrison, A. P. Nutman, and C.R.L. Friend. 1996. Evidence for life on Earth before 3,800 million years ago. *Nature* 384: 55–59. (Advances the case, now disputed, for an isotopic biosignature in very old rocks.)

Mojzsis, S. J., T. M. Harrison, and R. T. Pidgeon. 2001. Oxygen-isotope evidence from ancient zircons for liquid water at the Earth's surface 4,300 Myr ago. *Nature* 409: 178–181. (Insights into the early earth, based on the chemistry of mineral grains in ancient sandstone.)

Rasmussen, B. 2000. Filamentous microfossils in a 3,235-million-year-old volcanogenic massive sulphide. *Nature* 405: 676–679. (Perhaps the oldest convincing fossils of microorganisms.)

Rosing, M. T., 1999. C-13-depleted carbon microparticles in > 3700-Ma sea-floor sedimentary rocks from west Greenland. *Science* 283: 674–676. (An ancient isotopic record.)

Schopf, J. W., editor. 1983. *Earth's Earliest Biosphere*. Princeton University Press, Princeton, N.J., 543 pp. (A gold mine of information on the early Earth and life—dated but stimulating.)

Van Zuilen, M., A. Lepland, and G. Arrhenius. 2002. Reassessing the evidence for the earliest traces of life. *Nature* 418: 627–630. (A second argument that putative biosignatures claimed in Akilia rocks by Mojzsis and colleagues—see above—actually originated by physical processes during metamorphism.)

Chapter 5. The Emergence of Life

Brack, A., editor. 1999. *The Molecular Origins of Life: Assembling Pieces of the Puzzle*. Cambridge University Press, Cambridge. (An excellent volume that gathers together essays by leading figures in origins-of-life research.)

Darwin, C. 1969. *The Life and Letters of Charles Darwin*, volume 3. Johnson Reprint Corporation, New York. Originally published in 1887 by J. Murray, London. (Published source of Darwin's letter to Hooker.)

Darwin, E. 1804. *The Temple of Nature*. Reprinted by Pergamon, Elmsford, New York. (The elder Darwin's poetic outline of life's origin and evolution.)

Fry, I. 2000. *The Emergence of Life on Earth: A Historical and Scientific Overview*. Rutgers University Press, New Brunswick, N.J. (The best available one-volume account of life's origins, by a philosopher of science.)

Gilbert, W. 1986. The RNA world. *Nature* 319: 618. (A brief but influential essay on the implications of RNA enzymes for the origin of life.)

James, K. D., and A. D. Ellington. 1995. The search for missing links between self-replicating nucleic acides and the RNA world. *Origins of Life and Evolution of the Biosphere* 25: 515–530. (Thoughtful commentary on precursors to the RNA world.)

Joyce, G. F. 2002. The antiquity of RNA-based evolution. *Nature* 418: 214–221. (An excellent review of RNA's role in the emergence of life.)

Lee, D. H., J. R. Granja, J. A. Martinez, Kay Severin, and M. R. Ghadiri. 1996. A self-replicating peptide. *Nature* 382: 525–528. (Presents experimental evidence that self-replication by simple protein-like molecules may figure in the origin of life.)

Miller, S. L. 1953. A production of amino acids under possible primitive Earth conditions. *Science* 117: 527–528. (The pioneering experiment in origins-of-life research.)

Orgel, L. E. 1994. The origin of life on the Earth. *Scientific American* 271 (10): 77–83. (An authoritative primer on chemical evolution.)

Pace, N., and T. Marsh. 1986. RNA catalysis and the origin of life. *Origins of Life*

16: 97–116. (A good treatment of the discovery of ribozymes and its implications for prebiotic and early biological evolution.)

Wächtershäuser, G. 1992. Groundwork for an evolutionary biochemistry: The iron-sulphur world. *Progress in Biophysics and Molecular Biology* 58: 85–201. (A detailed statement of Wächtershäuser's metabolism-first view that life originated in hydrothermal vent systems.)

Szostak, J. W., D. P. Bartel, and P. L. Luisi. 2001. Synthesizing life. *Nature* 409: 387–390. (An introduction to experiments on directed evolution of catalytic RNA molecules.)

Chapter 6. The Oxygen Revolution

Key References on Gunflint and Other Late Archean/Early Proterozoic Paleontology

Amard, B., and J. Bertrand-Sarfati. 1997. Microfossils in 2000 My old cherty stromatolites of the Franceville Group, Gabon. *Precambrian Research* 81: 197–221.

Awarmik, S. M., and E. S. Barghoorn. 1977. The Gunflint microbiota. *Precambrian Research* 20: 357–374.

Barghoorn, E. S., and S. M. Tyler. 1965. Microfossils from the Gunflint chert. *Science* 147: 563–577.

Brocks J. J., G. A. Logan, R. Buick, and R. E. Summons. 1999. Archean molecular fossils and the early rise of eukaryotes. *Science* 285: 1033–1036.

Cloud, P. 1965. The significance of the Gunflint (Precambrian) microflora. *Science* 148: 27–35.

Golubic, S., and H. J. Hofmann. 1976. Comparison of Holocene and mid-Precambrian Entophysalidaceae (Cyanophyta) in stromatolitic algal mats: Cell division and degradation. *Journal of Paleontology* 50: 1074–1082.

Hofmann, H. J. 1976. Precambrian microflora, Belcher Islands, Canada: Significance and systematics. *Journal of Paleontology* 50: 1040–1073.

Knoll, A. H., E. S. Barghoorn, and S. M. Awramik. 1978. New organisms from the Aphebian Gunflint Iron Formation, Ontario. *Journal of Paleontology* 52: 976–992.

Knoll, A. H., P. K. Strother, and S. Rossi. 1988. Distribution and diagenesis of fossils from the lower Proterozoic Duck Creek Dolomite, Western Australia. *Precambrian Research* 38: 257–279.

Lanier, W. P. 1989. Interstitial and peloidal microfossils from the 2.0 Ga Gunflint Formation: Implications for the paleoecology of the Gunflint stromatolites. *Precambrian Research* 45: 291–318.

Simonson, B. M. 1985. Sedimentological constraints on the origins of Precambrian iron-formations. *Geological Society of America Bulletin* 96: 244–252.

Key References on the Early Proterozoic Oxygen Revolution

Canfield D. E. 1998. A new model for Proterozoic ocean chemistry. *Nature* 396: 450–453. (Articulates the hypothesis that iron formations disappeared because of growing hydrogen sulfide production in early Proterozoic oceans—a key paper in thinking about the biosphere's redox history.)

Catling, D. C., K. J. Zahnle, and C. P. McKay. 2001. Biogenic methane, hydrogen escape, and the irreversible oxidation of early Earth. *Science* 293: 839–843. (Proposes a solution to the mystery of why oxygen concentrations began to increase early in the Proterozoic Eon.)

Cloud, P. E. 1968. A working model of the primitive Earth. *American Journal of Science* 272: 537–548. (Key summary of traditional views on atmospheric history by one of the pioneers who developed them.)

Des Marais, D. J. 1997. See references to chapter 3.

Farquhar J., H. M. Bao, and M. Thiemens. 2000. Atmospheric influence of Earth's earliest sulfur cycle. *Science* 289: 756–758. (Introduces mass-independent fractionation of sulfur isotopes to the debate about atmospheric history.)

Habicht K. S., and D. E. Canfield. 1996. Sulphur isotope fractionation in modern microbial mats and the evolution of the sulphur cycle. *Nature* 382: 342–343. (A key paper that uses measurements of S-isotopic fractionation by living bacteria to constrain interpretations of the geochemical record.)

Ohmoto, H. 1996. Evidence in pre-2.2 Ga paleosols for the early evolution of atmospheric oxygen and terrestrial biotas. *Geology* 24: 1135–1138. (An approachable articulation of the minority view that oxygen was relatively abundant in Archean air and seawater.)

Rasmussen, B., and R. Buick. 1999. Redox state of the Archean atmosphere: Evidence from detrital heavy minerals in ca. 3250–2750 Ma sandstones from the Pilbara Craton, Australia. *Geology* 27: 115–118. (An important paper that documents detrital siderite and other minerals in late Archean sedimentary rocks, placing constraints on the oxygen content of the early atmosphere.)

Rye, R., and H. D. Holland. 1998. Paleosols and the evolution of atmospheric oxygen: A critical review. *American Journal of Science* 298: 621–672. (Reviews data on ancient weathering horizons, used by Holland and his collegaues to delve into atmospheric evolution.)

Chapter 7. The Cyanobacteria, Life's Microbial Heroes

Key References on Fossils and Geology of the Bil'yakh Group

Bartley, J. K., A. H. Knoll, J. P. Grotzinger, and V. N. Sergeev. 1999. Lithification and fabric genesis in precipitated stromatolites and associated peritidal dolomites, Mesoproterozoic Billyakh Group, Siberia. *SEPM Special Publication* 67: 59–74.

Golubic, S., V. N. Sergeev, and A. H. Knoll. 1995. Mesoproterozoic *Archaeoellipsoides*: Akinetes of heterocystous cyanobacteria. *Lethaia* 28: 285–298.

Knoll, A. H., and M. A. Semikhatov. 1998. The genesis and time distribution of two distinct Proterozoic stromatolite microstructures. *Palaios* 13: 408–422.

Sergeev, V. N., A. H. Knoll, and J. P. Grotzinger. 1995. Paleobiology of the Mesoproterozoic Billyakh Group, Anabar Uplift, northern Siberia. *Paleontological Society Memoir* 39, 37 pp.

Veis, A. F., and N. G. Vorbyeva. 1992. Riphean and Vendian microfossils of the Anabar Uplift. *Izvestia RAN, Seria geologocheskaya* 1: 114–130. (In Russian.)

Selected References on Cyanobacteria and Stromatolites

Giovannoni, S. J., S. Turner, G. L. Olsen, S. Barns, D. J. Lane, and N. R. Pace. 1988. Evolutionary relationships among cyanobacteria and green chloroplasts. *Journal of Bacteriology* 170: 3584–3692. (An important paper in which molecular sequence data are used to infer evolutionary relationships among cyanobacteria.)

Golubic, S. 1973. The relationship between blue-green algae and carbonate deposits, pp. 434–472 in N. G. Carr and B. A. Whitton, editors, *The Biology of Blue-Green Algae*. Oxford University Press, Oxford. (Basic reading for those interested in how cyanobacteria affect carbonate rocks and vice versa.)

Grotzinger, J. P., and A. H. Knoll. 1999. See references to chapter 3.

Knoll, A. H., and S. Golubic. 1992. Living and fossil cyanobacteria, pp. 450–462 in M. Schidlowski, S. Golubic, M. M. Kimberley, and P. A. Trudinger, editors, *Early Organic Evolution: Implications for Mineral and Energy Resources*. Springer-Verlag, Berlin. (Summarizes how detailed comparisons of fossil and living cyanobcteria have bolstered our understanding of cyanobacterial evolution.)

Lenski, R., and M. Travasiano. 1994. Dyanmics of adaptation and diversification: A 10,000 generation experiment with bacterial populations. *Proceedings of the National Academy of Sciences, USA* 91: 6808–6814. (A key paper exploring the tempo of bacterial evolution in a long-term laboratory experiment.)

Niklas, K. J. 1994. Morphological evolution through complex domains of fitness. *Proceedings of the National Academy of Sciences, USA* 91: 6772–6779. (A thoughtful inquiry into the reasons why some adaptive landscapes are steep while others are relatively smooth.)

Province, W. B. 1986. *Sewall Wright and Evolutionary Biology*. University of Chicago Press, Chicago, IL. (Wright and his work, including the concept of adaptive landscapes.)

Raaben, M. E., and M. A. Semikhatov. 1994. Dynamics of the global diversity of the suprageneric groupings of Proterozoic stromatolites. *Doklady, Russian Academy of Sciences* 349: 234–238. (A comprehensive synthesis of stromatolite diversity through time—best read as a "one, two, many" quantification unless you think you know what a "species" of stromatolite really is.)

Schopf, J. W. 1968. Microflora of the Bitter Springs Formation, late Precambrian, central Australia. *Journal of Paleontology* 42: 651–688. (A pioneering documentation of cyanobacteria in Proterozoic cherts.)

Walter, M. R. 1994. Stromatolites: The main source of information on the evolution of the early benthos, pp. 270–286 in S. Bengtson, editor, *Early Life on Earth*.

Columbia Univeristy Press, New York. (A personal update of the classic 1976 volume edited by Walter—see chapter 3 references.)

Whitton, B. A., and M. Potts, editors. 2000. *The Ecology of Cyanobacteria: Their Diversity in Time and Space.* Kluwer Academic Publishers, Dordrecht, Netherlands. (An up-to-date guide to cyanobacterial biology.)

Chapter 8. The Origins of Eukaryotic Cells

Key References on Eukaryotic Evolution and Phylogeny

Baldauf, S. L., A. J. Roger, I. Wenk-Siefert, and W. F. Doolittle. 2000. A kingdom-level phylogeny of eukaryotes based on combined protein data. *Science* 290: 972–977. (The best available guide to the genealogical relationships of eukaryotic organisms.)

Bui, E.T.N., P. J. Bradley, and P. J. Johnson. 1996. A common evolutionary origin for mitochondria and hydrogenosomes. *Proceedings of the National Academy of Sciences,USA* 93: 9651–9656. (A key paper relating hydrogenosomes to mitochondria and proteobacteria.)

Clark, C. G., and A. J. Roger. 1995. Direct evidence for secondary loss of mitochondria in *Entamoeba histolytica*. *Proceedings of the National Academy of Sciences, USA* 92: 6518–6521. (This paper launched research on mitochondrially derived genes in the nuclei of eukaryotes without mitochondria. As such, it fundamentally changed how we view early eukaryotic evolution.)

Delwiche, C. F. 1999. Tracing the thread of plastid diversity through the tapestry of life. *American Naturalist* 154: S164–S177. (An excellent summary of how photosynthesis spread through the Eucarya by primary, secondary, and tertiary endosymbioses.)

Douglas, S., and 9 others. 2001. The highly reduced genome of an enslaved algal nucleus. *Nature* 410: 1091–1096. (Genomic approach to the interplay between host and symbiont in the establishment of chloroplasts in a eukaryotic lineage.)

Dyer, B. D., and R. A. Oban. 1994. *Tracing the History of Eukaryotic Cells: The Enigmatic Smile.* Columbia University Press, New York. (Slightly dated but engaging introduction to the ideas of Lynn Margulis.)

Embley, T. M., and R. P. Hirt. 1998. Early branching eukaryotes? *Current Opinion in Genetics and Development* 8: 624–629. (A readable summary of recent work on eukaryotic phylogeny, including discussion of—and numerous references to—research on mitochondrial genes in the nuclear genomes of nonmitochondrial eukaryotes.)

Hartman, H., and A. Federov. 2002. The origin of the eukaryotic cell: A genomic investigation. *Proceedings of the National Academy of Sciences, USA* 99: 1420–1425. (Argues that genes found uniquely in eukaryotes record a third partner in the primordial symbiosis by which eukaryotic cells arose.)

Khakhina, L. N. 1992. *Concepts of Symbiogenesis. A Historical and Critical Account of the Research of Russian Botanists.* Yale University Press, New Haven, Conn. (An introduction to the work of Merezhkovsky and other early proponents of the endosymbiotic hypothesis.)

Margulis, L. 1981. *Symbiosis in Cell Evolution.* W. H. Freeman, San Francisco. (The classic statement of Margulis's views, updated and reissued in 1993.)

Martin, W., and M. Müller. 1998. The hydrogen hypothesis for the first eukaryote. *Nature* 392: 37–41. (Martin and Müller's provocative hypothesis, challenging but stimulating.)

Moreira, D., and P. López-Garcia. 1998. Symbiosis between methanogenic Archaea and δ-Proteobacteria as the origin of eukaryotes: The syntrophic hypothesis. *Journal of Molecular Evolution* 47: 517–530. (An independent articulation of the idea that eukaryotic biology reflects an early symbiosis between archeans and proteobacteria; differs in detail from the Martin-Müller hypothesis.)

Palmer, J. D. 1997. Organelle genomes: Going, going, gone! *Nature* 275: 790–791. (A readable review (with references) of research supporting the hypothesis that hydrogenosomes are mitochondria-like endosymbionts that lost all their genes.)

Roger, A. J. 1999. Reconstructing early events in eukaryotic evolution. *The American Naturalist* 154, supplement: S146–S163. (A good summary of molecular approaches to eukaryotic evolution.)

Sagan, L. 1967. On the origin of mitosing cells. *Journal of Theoretical Biology* 14: 225–274. (Lynn Margulis's—as Lynn Sagan—original essay on the endosymbiotic origins of mitochondria and chloroplasts. Controversial at the time, but now regarded as a classic.)

Sogin, M. 1997. History assignment: When was the mitochondrion founded? *Current Opinion in Genetics and Development* 7: 792–799. (Stimulating essay on alternative hypotheses for the origin of mitochondria in eukaryotic cells.)

Sogin, M. L., J. H. Gunderson, H. J. Elwood, R. A. Alonso, and D. A. Peattie. 1989. Phylogenetic meaning of the kingdom concept—an unusual ribosomal-RNA from *Giardia lamblia. Science* 243: 75–77. (The classic discussion of eukaryotic phylogeny as inferred from small-subunit ribosomal RNA genes.)

Thomas, L. 1979. *The Medusa and the Snail.* Viking Press, New York. (Source of Thomas's clever quote about committees and life.)

Chapter 9. Fossils of Early Eukaryotes

Key References on Doushantuo Geology and Paleontology

Most of the papers listed here were written in English. These publications will provide the interested reader with references to the extensive Chinese literature.

Barfod, G. H., F. Albarède, A. H. Knoll, S. Xiao, J. Baker, and R. Frei. 2002. New Lu-Hf and Pb-Pb age constraints on the earliest animal fossils. *Earth and Planetary Science Letters* 201: 203–212.

Chen, M., and Z. Zhao. 1992. Macrofossils from upper Doushantuo Formation in eastern Yangtze Gorges, China. *Acta Palaeontologica Sinica* 31: 513–529.

Li, C.-W., J.-Y. Chen, and T.-E. Hua. 1998. Precambrian sponges with cellular structures. *Science* 279: 879–882.

Steiner, M. 1994. Die neoproterozoischen Megalgen Südchinas. *Berliner geowissenschaftliche Abhandlungen (E)* 15: 1–146.

Xiao, S., and A. H. Knoll. 2000a. Phosphatized animal embryos from the Neoproterozoic Doushantuo Formation at Weng'an, Guizhou, South China. *Journal of Paleontology* 74: 767–788.

Xiao, S., and A. H. Knoll. 2000b. Eumetazoan fossils in terminal Proterozoic phosphorites? *Proceedings of the National Academy of Sciences, USA* 97: 13684–13689.

Xiao, S., M. Yuan, and A. H. Knoll. 1998. Morphological reconstruction *of Maiohephyton bifurcatum*, a possible brown alga from the Doushantuo Formation (Neoproterozoic), South China, and its implicatiuons for stramenopile evolution. *Journal of Paleontology* 72: 1072–1086.

Xiao, S., X. Yuan, M. Steiner, and A. H. Knoll. 2002. Carbonaceous macrofossils in a terminal Proterozoic shale: A systematic reassessment of the Miaohe biota, South China. *Journal of Paleontology* 76: 347–376.

Yuan, X., and H. J. Hofmann. 1998. New microfossils from the Neoproterozoic (Sinian) Doushantuo Formation, Weng'an, Guizhou Province, southwestern China. *Alcheringa* 22: 189–222.

Yuan, X., S. Xiao, L. Yin, A. H. Knoll, C. Zhao, and X. Mu. 2002. *Doushantuo Fossils: Life on the Eve of Animal Radiation.* University of Science and Technology of China Press, Beijing. (Written in Chinese, but worth seeking out for its many color photos of Doushantuo fossils.)

Zhang, Y. 1989. Multicellular thallophytes with differentiated tissues from late Proterozoic phosphate rocks of South China. *Lethaia* 22: 113–132.

Zhang, Y., L. Yin, S. Xiao, and A. H. Knoll. 1998. Permineralized fossils from the terminal Proterozoic Doushantuo Formation, South China. *Paleontological Society Memoir* 50: 1–52.

Key references on the biology of Proterozoic eukaryotes

Anbar, A., and A. H. Knoll. 2002. Proterozoic ocean chemistry and evolution: A bioinorganic bridge? *Science,* 297: 1137–1142. (Explores the consequences of sulfidic deep oceans for trace element chemistry, primary production, and, hence, algal evolution in Proterozoic oceans.)

Butterfield, N. J. 2000. *Bangiomorpha pubescens* n. gen., n. sp; implications for the evolution of sex, multicellularity, and the Mesoproterozoic/Neoproterozoic radiation of eukaryotes. *Paleobiology* 26: 386–404. (The oldest eukaryotic fossils that can be assigned to an extant algal clade.)

Butterfield, N. J., A. H. Knoll, and K. Swett. 1994. See references for chapter 3.

Fedonkin, M. A., and E. L. Yochelson. 2002. Middle Proterozoic (1.5 Ga) *Horodyskia moniliformis* Yochelson and Fedonkin, the oldest known tissue-

grade colonial eucaryote. *Smithsonian Contributions to Paleobiology* 94: 1–29. (Documents "string-of-beads" macrofossils in mid-Proterozoic rocks from North America.)

German, T. N. 1990. *Organic world one billion years ago.* Nauka, Leningrad, 52 pp. (A well-illustrated, bilingual guide to early eukaryotic fossils in Siberian rocks.)

Grey, K., and I. R. Williams. 1990. Problematic bedding-plane markings from the middle Proterozoic Manganese Subgroup, Bangemall Basin, Western Australia. *Precambrian Research* 46: 307–327. (A well-illustrated account of the macroscopic "string-of-beads" fossils impressed into mid-Proterozoic sandstone bedding surfaces.)

Hofmann, H. J. 1994. Problematic carbonaceous compressions ("metaphytes" and "worms"), pp. 342–358 in S. Bengtson, editor, *Early Life on Earth.* Columbia University Press, New York. (An authoritative account of the macroscopic compressions found in Proterozoic rocks.)

Javaux, E., A. H. Knoll, and M. R. Walter. 2001. Ecological and morphological complexity in early eukaryotic ecosystems. *Nature* 412: 66–69. (Documents the nature and distribution of eukaryotic fossils in a mid-Proterozoic seaway.)

Knoll, A. H. 1994. Proterozoic and Early Cambrian protists: Evidence for accelerating evolutionary tempo. *Proceedings of the National Academy of Sciences, USA* 91: 6743–6750. (Documents the rise in diversity and accelerating tempo of evolution in late Proterozoic eukaryotes.)

Porter, S. M., and A. H. Knoll. 2000. Testate amoebae in the Neoproterozoic Era: Evidence from vase-shaped microfossils in the Chuar Group, Grand Canyon. *Paleobiology* 26: 360–385. (Demonstrates the relationship between vase-shaped microfossils and living protozoans called testate amoebas.)

Shen, Y., D. E., Canfield, and A. H. Knoll. 2002. The chemistry of mid-Proterozoic oceans: Evidence from the McArthur Basin, northern Australia. *American Journal of Science* 302: 81–109. (Provides geochemical evidence of sulfidic deep waters in 1,730–1,640-million-year-old ocean basins.)

Summons, R. E., S. C. Brassell, G. Eglinton, E. Eaavans, R. J. Horodyski, N. Robinson, and D. M. Ward. 1988. Distinctive hydrocarbon biomarkers from fossiliferous sediment of the late Proterozoic Walcott Member, Chuar Group, Grand Canyon, Arizona. *Geochimica et Cosmochimica Acta* 52: 2625–2637. (Important study of eukaryotic biomarker molecules in late Proterozoic rocks.)

Swift, J. 1733. From *Poetry, A Rhapsody*, reprinted in *Bartlett's Familiar Quotations*, tenth edition (1919). Little, Brown, Boston. (Published source of Swift's famous verse.)

Vidal, G. 1976. Late Precambrian microfossils from the Visingsö beds in southern Sweden. *Fossils and Strata* 9: 1–57. (The paper that ignited global interest in Proterozoic biostratigraphy.)

Vidal, G., and M. Moczydlowska Vidal. 1997. Biodiversity, speciation, and extinction trends of Proterozoic and Cambrian phytoplankton. *Paleobiology* 23:

230–246. (Another view of early eukaryotic evolution; complements Knoll 1994.)

Zang, W., and M. R. Walter. 1992. Late Proterozoic and Cambrian microfossils and biostratigraphy, Amadeus Basin, central Australia. *Association of Australasian Palaeontologists Memoir* 12: 1–132. (A magnificent discovery of microfossils much like those in Doushantuo cherts and phosphates, beautifully illustrated, if a bit overenthusiastic in naming new species.)

Chapter 10. Animals Take the Stage

Key References on Nama Geology and Paleontology

Droser, M. L., S. Jensen, and J. G. Gehling. 2002. Trace fossils and substrates of the terminal Proterozoic-Cambrian transition: Implications for the record of early bilaterians and sediment mixing. *Proceedings of the National Academy of Sciences, USA* 99: 12572–12576.

Germs, G.J.B. 1972. New shelly fossils from the Nama Group, Namibia (South West Africa). *American Journal of Science* 272: 752–761.

Germs, G.J.B., A. H. Knoll, and G. Vidal. 1986. Latest Proterozoic microfossils from the Nama Group, Namibia (South West Africa). *Precambrian Research* 73: 137–151.

Grant, S.W.F. 1990. Shell structure and distribution of *Cloudina*, a potential index fossil for the terminal Proterozoic. *American Journal of Science* 290A: 261–294.

Grotzinger, J. P., S. A. Bowring, B. Z. Saylor, and A. J. Kaufman. 1995. Biostratigraphic and geochronologic constraints on early animal evolution. *Science* 270: 598–604.

Grotzinger, J. P., W. A. Watters, and A. H. Knoll. 2000. Calcified metazoans in thrombolite-stromatolite reefs of the terminal Proterozoic Nama Group, Namibia. *Paleobiology* 26: 334–359.

Gürich, G. 1933. Die Kuibis-Fossilien der Nama Formation von Südwest-Afrika. *Paläontologische Zeitschrift* 15: 137–154.

Narbonne, G. M., B. Z. Saylor, and J. P. Grotzinger. 1997. The youngest Ediacaran fossils from southern Africa. *Journal of Paleontology* 71: 953–967.

Pflug, H. D. 1970, 1970, 1972. Zur Fauna der nama-Schichten in Südwest Afrika. I. Pteridinia, Bau und systematische Zugehörigkeit. *Palaeontolographica Abteilung A* 134: 226–262; II. Rangidae, Bau und systematische Zugehörigkeit. *Palaeontolographica Abteilung A* 135: 198–231; III. Erniettomorpha, Bau und systematische Zugehörigkeit. *Palaeontolographica Abteilung A* 139: 134–170.

Wood, R. A., J. P. Grotzinger, and J.A.D. Dickson. 2002. Proterozoic modular biomineralized metazoan from the Nama Group. *Science* 296: 2383–2386.

General References on Ediacaran Fossils and Their Interpretation

Buss, L. W., and A. Seilacher. 1994. The phylum Vendobionta: A sister group of the eumetazoa? *Paleobiology* 20: 1–4. (A stimulating essay on the phylogenetic placement of vendobionts—a "limited modified withdrawal" from the extinct kingdom hypothesis.)

Fedonkin, M. A. 1990. Systematic description of the Vendian metazoa, pp. 71–120 in B. S. Sokolov and A. B. Iwanowski, editors, *The Vendian System,* volume I. Springer-Verlag, Berlin. (The best English-language summary of Fedonkin's groundbreaking research on Ediacaran fossils from the White Sea, Russia.)

Fedonkin, M. A., and B. M. Waggoner. 1997. The late Precambrian fossil *Kimberella* is a mollusc-like bilaterian organism. *Nature* 388: 868–871. (An important study, based on new fossils from the White Sea, that links an Ediacaran species to a bilaterian stem group.)

Gehling, J. M. 1999. Microbial mats in terminal Proterozoic siliciclastics: Ediacaran death masks. *Palaios* 14: 40–57. (Offers the best available answer to the question of how Ediacaran animals became fossilized.)

Gehling, J. G., G. M. Narbonne, and M. M. Anderson. 2000. The first named Ediacaran body fossil, *Aspidella terranovica. Palaeontology* 43: 427–456. (A good summary of discoidal fossils in Ediacaran assemblages and their biological interpretation.)

Glaessner, M. F. 1983. *The Dawn of Animal Life: A Biohistorical Study.* Cambridge University Press, Cambridge, 244 pp. (The valedictory statement by Ediacaran paleontology's original master; summarizes Glaessner's research on Australian and Namibian fossils.)

Jenkins, R.J.F. 1992. Functional and ecological aspects of Ediacaran assemblages, pp. 131–176 in J. H. Lipps and P. W. Signor, editors, *Origin and Evolution of the Metazoa.* Plenum, New York. (Key contribution from another leading interpreter of Ediacaran fossils.)

Narbonne, G. M. 1998. The Ediacara biota: A terminal Proterozoic experiment in the evolution of life. *GSA Today* 8 (2): 1–7. (This primer provides a good starting point for students of Ediacaran paleontology.)

Runnegar, B. 1995. Vendobionta or Metazoa? Developments in understanding the Ediacara "fauna." *Neues Jahrbuch für Geologie und Paläontologie, Abhandlungen* 195: 303–318. (A thoughtful essay on the form, function, and biological relationships of Ediacaran fossils.)

Seilacher, A. 1992. Vendobionta and psammocorallia: Lost constructions of Precambrian evolution. *Geological Society of London Journal* 149: 607–613. (A mature statement of Seilacher's stimulating and controversial interpretations.)

Chapter 11. Cambrian Redux

Bengtson, S. 1994. The advent of animal skeletons, pp. 414–425 in S. Bengtson, editor, *Early Life on Earth.* Nobel Symposium 84, Columbia University Press, New York. (Wise commentary on predation and the Cambrian evolution of mineralized skeletons.)

Bengtson, S., S. Conway Morris, B. J. Cooper, P. A. Jell, and B. N. Runnegar. 1990. Early Cambrian fossils from South Australia. *Memoirs of the Association of*

Australasian Paleontologists 9: 1–364. (An exceptionally fine treatment of Cambrian small shelly fossils.)

Bowring, S. A., and D. H. Erwin. 1998. A new look at evolutionary rates in deep time: Uniting paleontology and high-precision geochronology. *GSA Today* 8 (9): 1–8. (Lays out the modern timescale for early animal diversification.)

Budd, G. E., and S. Jensen. 2000. A critical reappraisal of the fossil record of the bilaterian phyla. *Biological Reviews* 75: 253–295. (A lawyerly but valuable essay on Cambrian animals, stressing that most are stem members of bilaterian phyla or classes.)

Carroll, S. B., J. K. Grenier, and S. C. Weatherbee. 2001. *From DNA to Diversity.* Blackwell Scientific, Oxford. (A superb introduction to developmental genetics and animal evolution.)

Chen, J., and G. Zhou. 1997. Biology of the Chenjiang fauna. *Bulletin of the National Museum of Natural Science (Taiwan)* 10: 11–106. (A well-illustrated account of Burgess's older brother.)

Conan Doyle, A. 1892. *Silver Blaze,* reprinted in *Complete Sherlock Holmes.* Doubleday, New York, 1960. (Source of the quote about the dog that did not bark.)

Conway Morris, S. 1998. See references to chapter 1.

Davidson, E. H. 2001. *Genomic Regulatory Systems: Development and Evolution.* Academic Press, New York. (A masterful advanced account of developmental genetics and its evolutionary consequences. Includes discussion of Davidson's stimulating and controversial hypothesis that early animals looked much like living larvae and only with the evolution of "set-aside cells" gained the capacity for large complex bodies.)

Eliott, T. S. 1942. Little Gidding, in *The Complete Poems and Plays, 1909–1950.* Harcourt Brace and World, New York. (Excerpt from "Little Gidding" in Four Quartets, copyright 1942 by T. S. Eliot and renewed 1970 by Esme Valerie Eliot, reprinted by permission of Harcourt, Inc.)

Fortey, R. A., D.E.G. Briggs, and M. A. Wills. 1996. The Cambrian evolutionary 'Explosion': Decoupling cladogenesis from morphological disparity. *Biological Journal of the Linnaean Society* 57: 13–33. (Makes the case that animal lineages may have diverged earlier but evolved their characteristic body plans only in the Cambrian.)

Garey, J. R., and A. Schmidt-Rhaesa. 1998. The essential role of "minor" phyla in molecular studies of animal evolution. *American Zoologist* 38: 907–917. (Simultaneously a ringing defense of the "little guys," phyla of small animals seldom included in phylogenetic analyses, and a good discussion of animal phylogeny as revealed by molecular data.)

Gould, S. J. 1989. *Wonderful Life: The Burgess Shale and the Nature of History.* Norton, New York. (A page-turner that provides elegant descriptions of Burgess animals, with stimulating if controversial interpretations of their paleobiology.)

Jensen, S. 1992. Trace fossils from the lower Cambrian Mickwitzia sandstone, south-central Sweden. *Fossils and Strata* 42: 1–111. (A good place to learn about trace fossils across the Proterozoic-Cambrian boundary.)

Knoll, A. H., and S. B. Carroll. 1999. Early animal evolution: Emerging views from comparative biology and geology. *Science* 284: 2129–2137. (An attempt to integrate insights from developmental genetics and paleontology—Cliff Notes for this chapter.)

Miklos, G.L.G. 1993. Emergence of organizational complexities during metazoan evolution: Perspectives from molecular biology, palaeontology, and neo-Darwinism. *Association of Australasian Palaeontologists, Memoir* 15: 7–41. (Stresses the importance of emergent properties of evolving organic systems, especially the nervous system of early animals.)

Ruppert, E. E., and R. D. Barnes. 1994. *Invertebrate Zoology*, sixth edition. Saunders College Publishing, Fort Worth. (A good place to learn about the tremendous diversity of animal life.)

Smith, A. 1999. Dating the origins of metazoan body plans. *Evolution and Development* 1: 138–142. (A thoughtful attempt to reconcile molecular clock and geological dates for early animal divergence.)

Valentine, J. W., D. Jablonski, and D. H. Erwin. 1999. Fossils, molecules and embryos: New perspectives on the Cambrian Explosion. *Development* 126: 851–859. (Another view on how genetics and paleontology may fit together in studies of early animal evolution.)

Wray, G. A., J. S. Levinton, and L. H. Shapiro. 1996. Molecular evidence for deep Precambrian divergences among metazoan phyla. *Science* 274: 568–573. (Uses molecular clocks to argue for early animal diversification.)

Chapter 12. Dynamic Earth, Permissive Ecology

Key References on Late Proterozoic Ice Ages

Evans, D.A.D. 2000. Stratigraphic, geochronological, and paleomagnetic constraints upon the Neoproterozoic climatic paradox. *American Journal of Science* 300: 347–433.

Harland, W. B., and M. S. Rudwick. 1964. The great Infra–Cambrian ice age. *Scientific American* 211 (2): 28–36.

Hoffman, P. F. 1999. The break-up of Rodinia, birth of Gondwana, true polar wander, and the Snowball Earth. *Journal of African Earth Sciences* 28: 17–33.

Hoffman, P. F., A. J. Kaufman, G. P. Halverson, and D. P. Schrag. 1998. A Neoproterozoic Snowball Earth. *Science* 281: 1342–1346.

Hoffman, P. F., and D. P. Schrag. 2002. The snowball Earth hypothesis: testing the limits of global change. *Terra Nova* 14: 129–155.

Hyde, W. T., T. J. Crowley, S. K. Baum, and W. R. Peltier. 2000. Neoproterozoic 'Snowball Earth' simulations with a coupled climate/ice sheet model. *Nature* 405: 425–429.

Kennedy, M. J., N. Christie-Blick, and A. R. Prave. 2001. Carbon isotopic composition of Neoproterozoic glacial carbonates as a rest of paleoceanographic models for Snowball Earth phenomena. *Geology* 29: 1135–1138.

Kirschvink, J. 1992. Late Proterozoic low latitude glaciation: The Snowball Earth, pp. 51–52 in J. W. Schopf and C. Klein, editors, *The Proterozoic Biosphere: A Multidisciplinary Study*. Cambridge University Press, Cambridge.

Schrag, D. P. , R. A. Berner, P. F. Hoffman, and G. P. Halverson. 2002. On the initiation of a snowball Earth. *Geochemistry Geophysics Geosystems* 3: art. no. 1036. (Electronic journal file, accessible by Internet.)

Vidal, G., and A. H. Knoll. 1982. Radiations and extinction of plankton in the late Proterozoic and Early Cambrian. *Nature* 297: 57–60.

References on Animals and Late Proterozoic Oxygen Increase

Canfield, D. E., and A. Teske. 1996. Late Proterozoic rise in atmospheric oxygen concentration inferred from phlyogenetic and sulphur-isotope studies. *Nature* 382: 127–132.

Derry, L. A., A. J. Kaufman, and S. B. Jacobsen. 1992. Sedimentary cycling and environmental change in the late Proterozoic: Evidence from stable and radiogenic isotopes. *Geochimica et Cosmochimica Acta* 56: 1317–1329.

Graham, J. B. 1988. Ecological and evolutionary aspects of integumentary respiration: Body size, diffusion, and the Invertebrata. *American Zoologist* 28: 1031–1045.

Knoll, A. H., J. M. Hayes, J. Kaufman, K. Swett, and I. Lambert. 1986. Secular variation in carbon isotope ratios from upper Proterozoic successions of Svalbard and East Greenland. *Nature* 321: 832–838.

Nursall, J. R. 1959. Oxygen as a prerequisite to the origin of the metazoa. *Nature* 183: 1170–1172.

Rhoads, D. C., and J. W. Morse. 1971. Evolutionary and ecological significance of oxygen-deficient marine basins. *Lethaia* 4: 413–428.

Runnegar, B. 1982. Oxygen requirements, biology and phylogenetic significance of the late Precambrian worm *Dickinsonia*, and the evolution of the burrowing habit. *Alcheringa* 6: 223–239.

Towe, K. M. 1970. Oxygen-collagen priority and the early metazoan fossil record. *Proceedings of the National Academy of Sciences, USA* 65: 781–788.

References on Proterozoic-Cambrian Boundary Perturbation and Its Biological Consequences

Amthor, J. E., J. P. Grotzinger, et al. 2003. Extinction of *Cloudina* and *Namacalathus* at the Precambrian-Cambrian boundary in Oman. *Geology* 31: 431–434.

Bartley, J. K., M. Pope, A. H. Knoll, M. A. Semikhatov, and P. Yu. Petrov. 1998. A Vendian-Cambrian boundary succession from the northwestern margin of the Siberian Platform: Stratigraphy, paleontology, chemostratigraphy, and correlation. *Geological Magazine* 135: 473–494.

Kimura, H., and Y. Watanabe. 2001. Oceanic anoxia at the Precambrian-Cambrian boundary. *Geology* 29: 995–998.

Knoll, A. H., and S. B. Carroll. 1999. See references to chapter 11.

Chapter 13. Paleontology ad Astra

Key References on the Mars Meteorite Debate

Barber, D. J., and E.R.D. Scott. 2002. Origin of supposedly biogenic magnetite in martian meteorite Allan Hills 84001. *Proceedings of the National Academy of Sciences, USA* 99: 6556–6561.

Bradley, J. P., R. P. Harvey, and H. Y. McSween, Jr. 1996. Magnetite whiskers and platelets in the ALH84001 Martian meteorite: Evidence of vapor phase growth. *Geochimica et Cosmochimica Acta* 60: 5149–5155.

Clemett, S. J., X.D.F. Chillier, S. Gillette, R. N. Zare, M. Maurette, C. Engrand, and G. Kurat. 1998. Observation of indigenous polycyclic aromatic hydrocarbons in "giant" carbonaceous antarctic micrometeorites. *Origins of Life and Evolution of the Biosphere* 28: 425–448.

Gibson, E. K., Jr., D. S. McKay, K. Thomas-Keprta, and C. S. Romanek. 1997. The case for relic life on Mars. *Scientific American* 277 (12): 58–65.

Golden, D. C., D. W. Ming, H. V. Lauer, Jr., C. S. Schwandt, R. V. Morris, G. E. Lofgren, and G. A. McKay. 2002. Inorganic formation of "truncated hexa-octahedral" magnetite: Implications for inorganic processes in Martian meteorite ALH-84001. *Abstracts, Lunar and Planetary Science Conference.*

Golden, D. C., D. W. Ming, C. S. Schwandt, H. V. Lauer, Jr., R. A. Socki, R. V. Morris, G. E. Lofgren, and G. A. McKay. 2001. A simple inorganic process for formation of carbonates, magnetite, and sulfides in Martian meteorite ALH84001. *American Mineralogist* 86: 370–375.

Kerr, R. A. 2002. See references for chapter 4.

McKay, D. S., E. K. Gibson, Jr., K. L. Thomas-Keprta, H. Vali, C. S. Romaneck, S. J. Clemett, X.D.F. Chillier, C. R. Maechling, and R. N. Zare. 1996. Search for past life on Mars: Possible relic biogenic activity in martian meteorite ALH84001. *Science* 273: 924–930.

Mittlefeldt, D. W. 1994. ALH84001, a cumulate orthopyroxenite member of the martian meteorite clan. *Meteoritics* 29: 214–221.

Thomas-Keprta, K. L., and 9 others. 2001. Truncated hexa-octahedral magnetite crystals in ALH84001: Presumptive biosignatures. *Proceedings of the National Academy of Sciences, USA* 98: 2164–2169.

Treiman, A. Recent scientific papers on ALH 84001 explained, with insightful and totally objective commentaries. http://cass.jsc.nasa.gov/lpi/meteorites/alhnpap.html (An excellent website that provides an objective guide to papers on the Allan Hills meteorite. Sadly, there are no updates after December 12, 2000.)

Selected References on Astrobiology and Astropaleontology

Carr, M. 1996. *Water on Mars*. Oxford University Press, Oxford. (A technical but authoritative account of the key requirement for martian life.)

Davies, P. 1995. *Are We Alone?* Penguin Books, London. (Commentary on the big question by an astute and articulate astrophysicist.)

Des Marais, D., editor. 1997. *The Pale Blue Dot Workshop: Spectroscopic Search for Life on Extrasolar Planets*. NASA Conference Publication 10154, 39 pp. (Results of a workshop on how to detect life in nearby solar systems, providing astrobiological justification for the Terrestrial Planet Finder.)

Farmer, J. D., and D. J. Des Marais. 1999. Exploring for a record of ancient martian life. *Journal of Geophysical Research* 104: 26,977–26,995. (Articulates a strategy for Mars astropaleontology.)

Goldsmith, D., and T. Owen. 2001. *The Search for Life in the Universe*, third edition. University Science Books, Sausalito, Calif. (A readable guide to the universe and the life it may hold.)

Gopnik, A. 2002. The porcupine: A pilgrimage to Popper. *The New Yorker*, April 1, 2002: 88—93. (A keenly observed essay on Karl Popper and the nature of scientific argument.)

Hesse, H. 1943. *The Glass Bead Game*. Reissue edition by Henry Holt, New York, 1990. (Source of the quote in chapter 13.)

Lissauer, J. J. 1999. How common are habitable planets? *Nature* 402: C11–C14. (Thoughtful essay on a difficult question.)

Lunine, J. I. 1999. In search of planets and life around other stars. *Proceedings of the National Academy of Sciences, USA* 96: 5353–5355. (A readable introduction to extrasolar planet detection, with thoughts on its implications for astrobiology.)

Malin, M. C., and K. S. Edgett. 2000. Evidence for recent groundwater seepage and surface runoff on Mars. *Science* 288: 2330–2335. (New observations suggesting that water was present relatively recently on the martian surface.)

McSween, H. Y., Jr. 1997. *Fanfare for Earth: The Origin of Our Planet and Life*. St. Martin's Press, New York. (An engaging portrait of our planetary history and its implications for life in the universe.)

Shostak, S., B. Jakosky, and J. O. Bennett. 2002. *Life in the Universe*. Addison-Wesley, Boston. (An excellent primer on all things astrobiological.)

Tarter, J. C., and C. F. Chyba. 1999. Is there life elsewhere in the universe? *Scientific American* 281 (12): 118–123. (SETI and the search for extraterrestrial life.)

Walter, M. R. 1999. *The Search for Life on Mars*. Perseus Books, Reading Mass. (A personal tour of the astrobiological landscape by one of Precambrian paleontology's ranking experts.)

Ward, P. D., and D. Brownlee. 2000. *Rare Earth: Why Complex Life Is Uncommon in the Universe*. Copernicus Books, New York. (A paleontologist and a space scientist team up to argue exactly what their title says.)

Epilogue

Barnes, J. 1984. *Flaubert's Parrot*. Jonathan Cape, London. (Source of my opening quote; reprinted with permission.)

Bradie, M. 1994. The Secret Chain: Evolution and Ethics. State University of New York Press, Albany, N.Y. (Philosophy at the interface between evolution and human ethics.)

de Duve, C. 1995. *Vital Dust: Life as a Cosmic Imperative*. Basic Books, New York. (A stimulating guide to the history of life by a great cell biologist at ease with science and Catholicism.)

Gould, S. J. 1999. *Rocks of Ages: Science and Religion in the Fullness of Life*. Ballantine Books, New York. (Gould weighs in on science and religion as "non-overlapping magisteria"—endeavors with separate goals and practices.)

Myers, N., and A. H. Knoll, editors. 2001. The biotic crisis and the future of evolution. *Proceedings of the National Academy of Sciences, USA* 98: 5389–5480. (Papers from a colloquium on our evolutionary future.)

O'Hara, R. J. 1992. Telling the tree. *Biology and Philosophy* 7: 135–160. (A thoughtful essay on evolutionary narrative in the age of phylogeny.)

Ruse, M. 2001. *Can a Darwinian be a Christian*? Cambridge University Press, Cambridge. (A lively philosophical guide to the interface between scientific and Christian thought. Thoroughly recommended.)

Sproul, B. C. 1979. *Primal Myths: Creation Myths around the World*. HarperCollins, New York. (A guide to the richness of traditional thought on creation.)

Tucker, M. E., and J. A. Grim, editors. 2001. Religion and ecology: Can the climate change? *Daedalus* 130 (4): 1–306. (Fifteen essays on ethics and religion in an age of global change.)

Index